Distributed Network Structure Estimation Using Consensus Methods

Synthesis Lectures on Communications

Editor
William Tranter, *Virginia Tech*

Basic Simulation Models of Phase Tracking Devices Using MATLAB
William Tranter, Ratchaneekorn Thamvichai, and Tamal Bose
2010

Joint Source Channel Coding Using Arithmetic Codes
Dongsheng Bi, Michael W. Hoffman, and Khalid Sayood
2009

Fundamentals of Spread Spectrum Modulation
Rodger E. Ziemer
2007

Code Division Multiple Access(CDMA)
R. Michael Buehrer
2006

Game Theory for Wireless Engineers
Allen B. MacKenzie and Luiz A. DaSilva
2006

Distributed Network Structure Estimation using Consensus Methods
Sai Zhang, Cihan Tepedelenlioglu, Andreas Spanias, and Mahesh Banavar

ISBN: 978-3-031-00556-5   paperback
ISBN: 978-3-031-01684-4   ebook
ISBN: 978-3-031-00050-8   hardcover

DOI 10.1007/978-3-031-01684-4

A Publication in the Springer series
SYNTHESIS LECTURES ON COMMUNICATIONS

Lecture #13
Series Editor: William Tranter, *Virginia Tech*
Series ISSN
Print 1932-1244   Electronic 1932-1708

Distributed Network Structure Estimation Using Consensus Methods

Sai Zhang
Arizona State University

Cihan Tepedelenlioglu
Arizona State University

Andreas Spanias
Arizona State University

Mahesh Banavar
Clarkson University

SYNTHESIS LECTURES ON COMMUNICATIONS #13

ABSTRACT

The area of detection and estimation in a distributed wireless sensor network (WSN) has several applications, including military surveillance, sustainability, health monitoring, and Internet of Things (IoT). Compared with a wired centralized sensor network, a distributed WSN has many advantages including scalability and robustness to sensor node failures. In this book, we address the problem of estimating the structure of distributed WSNs. First, we provide a literature review in: (a) graph theory; (b) network area estimation; and (c) existing consensus algorithms, including average consensus and max consensus. Second, a distributed algorithm for counting the total number of nodes in a wireless sensor network with noisy communication channels is introduced. Then, a distributed network degree distribution estimation (DNDD) algorithm is described. The DNDD algorithm is based on average consensus and in-network empirical mass function estimation. Finally, a fully distributed algorithm for estimating the center and the coverage region of a wireless sensor network is described. The algorithms introduced are appropriate for most connected distributed networks. The performance of the algorithms is analyzed theoretically, and simulations are performed and presented to validate the theoretical results. In this book, we also describe how the introduced algorithms can be used to learn global data information and the global data region.

KEYWORDS

wireless sensor networks, diffusion adaptation, node counting, Internet-of-Things (IoT)

Contents

Preface

The sensor signal and information processing (SenSIP) center at Arizona State University is home to several sensor related research projects. This manuscript is intended to be an introduction to the topic of sensor network structure estimation which is based on the dissertation of the lead author. Earlier work associated with this topic begun in the group of Dr. Tepedelenlioglu and continued later by work done by all co-authors in an NSF FRP grant to the SenSIP I/UCRC site. This later gave rise to collaborative work between SenSIP researchers and Dr. Banavar at Clarkson University funded by a SenSIP I/UCRC grant on sensor networks. The specific focus of the book is based on Dr. Sai Zhang's dissertation which was completed under the supervision of Drs. Tepedelenlioglu and Spanias.

This book summarizes several sensor network methods with the focus on consensus-based estimation. The topics of distributed node counting, degree distribution estimation, and network center and radius estimation are discussed in detail. Several applications are also described in the introduction and at the end of the book with several papers cited for further reading.

Sai Zhang, Cihan Tepedelenlioglu, Andreas Spanias, and Mahesh Banavar
February 2018

Acknowledgments

The work in this manuscript was supported in part by the SenSIP Center, Arizona State University, the NCSS I/UCRC site, and NSF FRP award 123034.

Sai Zhang, Cihan Tepedelenlioglu, Andreas Spanias, and Mahesh Banavar
February 2018

CHAPTER 1

Introduction

1.1 WIRELESS SENSOR NETWORKS

A wireless sensor network (WSN) consists of a group of specialized distributed sensors used to monitor conditions such as temperature, pressure, sound vibration and pollutant levels [1–4]. The sensors are connected through wireless infrastructure.

In a centralized WSN, there is a central node (fusion center) controlling the entire network. The spatially distributed sensor nodes are used to monitor physical or environmental conditions and pass their data through the network to the fusion center [1, 5, 6]. An example of a WSN with a fusion center is given in Figure 1.1. In a centralized wireless sensor network, the fusion center has all the data from the sensor nodes. Therefore, functions of the data, such as the average, the maximum, or the minimum of the initial measurements can be easily computed at the fusion center. However, there are also disadvantages with using a centralized fusion center in that the entire network will collapse if the fusion center malfunctions. Moreover, since all sensors need to communicate with the fusion center, the bandwidth and power consumption are usually high [7].

In a distributed WSN without a fusion center, sensor nodes exchange data only with their neighbors [8–11]. An example of the distributed wireless sensor network without a fusion center is given in Figure 1.2. A WSN without a fusion center can function autonomously. Compared to the centralized network, a distributed network without a fusion center is more scalable than a centralized system and is also more robust to link failures [12]. Since the nodes in a decentralized network communicate only with their neighbors, the sensors require less power [13–17, 68] and can support higher data rates [18]. However, function computation in a distributed WSN is usually more complicated than in a centralized network. Moreover, convergence of the states of nodes is slow in a distributed sensor network [19–21].

1.2 APPLICATIONS

Wireless sensor networks are widely used in both military and industrial applications. A comprehensive review of wireless sensor network applications is given in [6, 22–24].

In military and defense systems, wireless sensor networks are often used for target tracking [25, 26]. In [26], a collaborative signal processing approach to WSNs is described where the tracking of multiple targets is presented. Improved moving vehicle target classification in battlefields using WSNs is introduced in [25]. In [27], a method to calculate the quality of a WSN

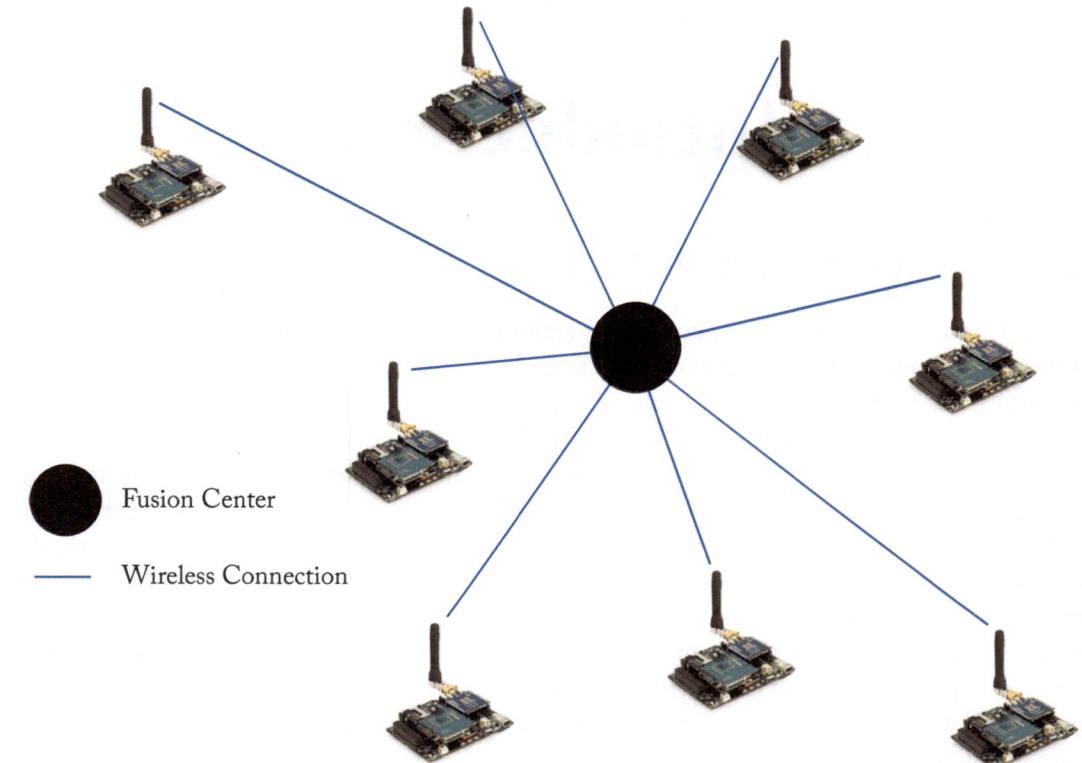

Figure 1.1: Centralized wireless sensor network framework.

based on network structure information such as the number of sensor nodes and the network coverage region is described. The method is used for area monitoring and border surveillance and control. WSNs used in border surveillance applications are also presented in [27, 28], where data from different types of sensors are gathered at the command center and used to initiate appropriate actions. An example of WSNs used for border monitoring and control is shown in Figure 1.3.

Wireless sensor networks are also widely used in industrial and commercial applications. Applications include indoor/outdoor sensing [29–31], environmental monitoring [23, 32–34], power monitoring [35], tracking a target or predicting the dynamics of a target [6, 36–38], network security [39], localization [40–42], Internet of things (IoT) [43] and detection and estimation [44–47]. Sensors powered by solar cells are described in [6], where they are used to protect forests without human intervention. WSNs are also useful in health monitoring and medical diagnosis, where they are used to sense and store physiological data [48–52]. In [53–55], sensors and cameras are used for measuring and recording parameters such as blood pres-

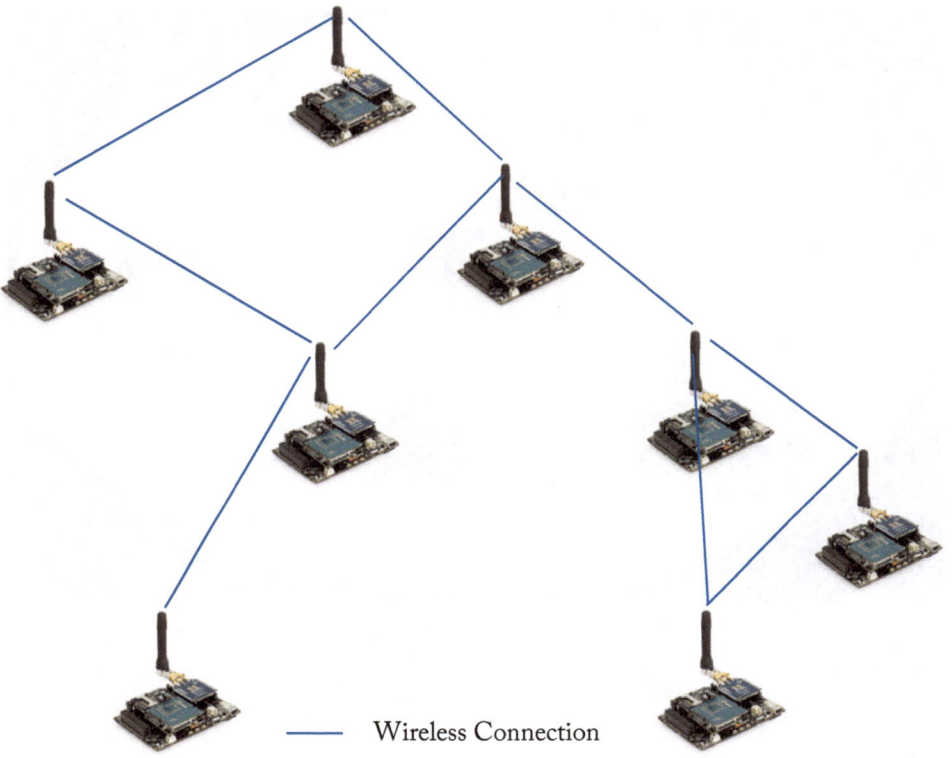

Figure 1.2: Distributed wireless sensor network framework without a fusion center.

sure and heart rhythm for real-time health monitoring. In [56–59], a solar monitoring facility equipped with sensors is used for solar array monitoring. Sensors are used for fault detection and for optimizing the efficiency of the solar array. The sensor-enabled solar facility behaves as an IoT cyber-physical system that can detect faults monitor irradiance and change connections to maximize power [24, 58]. Figure 1.4 summarizes the wide range of applications of WSNs.

1.3   CONSENSUS METHODS IN DISTRIBUTED WSNs

The consensus problem has a long history of research in the fields of computer and network science. It requires an agreement among a number of agents to derive a single data value [60, 61]. In WSN consensus, algorithms converge to a global agreement relying on local communications [62–65].

The problem of consensus estimation in WSNs has attracted great interest in recent years since it is useful in diverse applications, especially in network science [61], control [66, 67], and communications [64, 68–70]. A comprehensive review of the applications of consensus al-

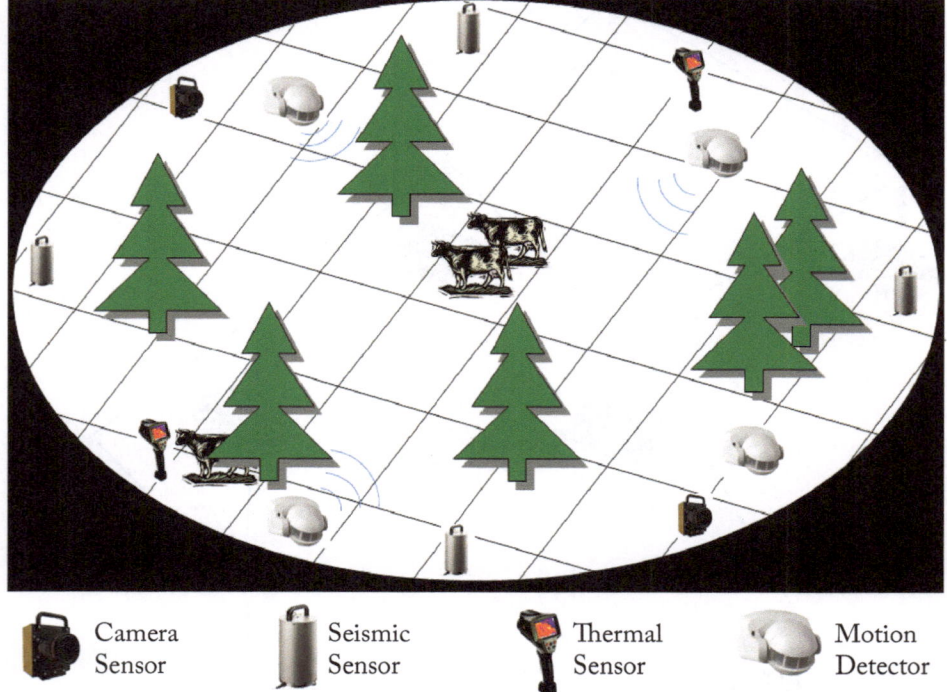

| Camera Sensor | Seismic Sensor | Thermal Sensor | Motion Detector |

Figure 1.3: WSNs for security and border control applications.

gorithms is given in [61], where applications, including synchronization of coupled oscillators [71], flocking for mobile agents [72] and distributed formation control [73, 74] are discussed. Average consensus can be used for distributed sensor fusion [75], distributed decision making [76], and distributed detection and estimation [8, 13, 14, 68, 75]. In [77], distributed time synchronization in wireless sensor networks is achieved by max consensus.

The most widely used consensus algorithm in literature is average consensus [69]. In average consensus problems, the states of nodes are designed to converge to the average or the sample mean of the initial measurements or states. The most basic linear average consensus is reported in [66, 69], and average consensus with communications noise between nodes and with link failures is considered in [64]. In [78, 79], nonlinear average consensus is considered and nonlinear functions are used to bound the transmission power and limit the variance of impulsive channel noise, respectively. A review on average consensus is given in Section 2.2.1.

Max consensus is also widely studied [7, 80] where the nodes converge to the maximum of the initial measurements. With perfect communication channels, the max operator is usually used for max consensus [66, 80–84]. During the iteration process, the state at each sensor node is updated by the largest value obtained. The problem of max consensus with noisy communication

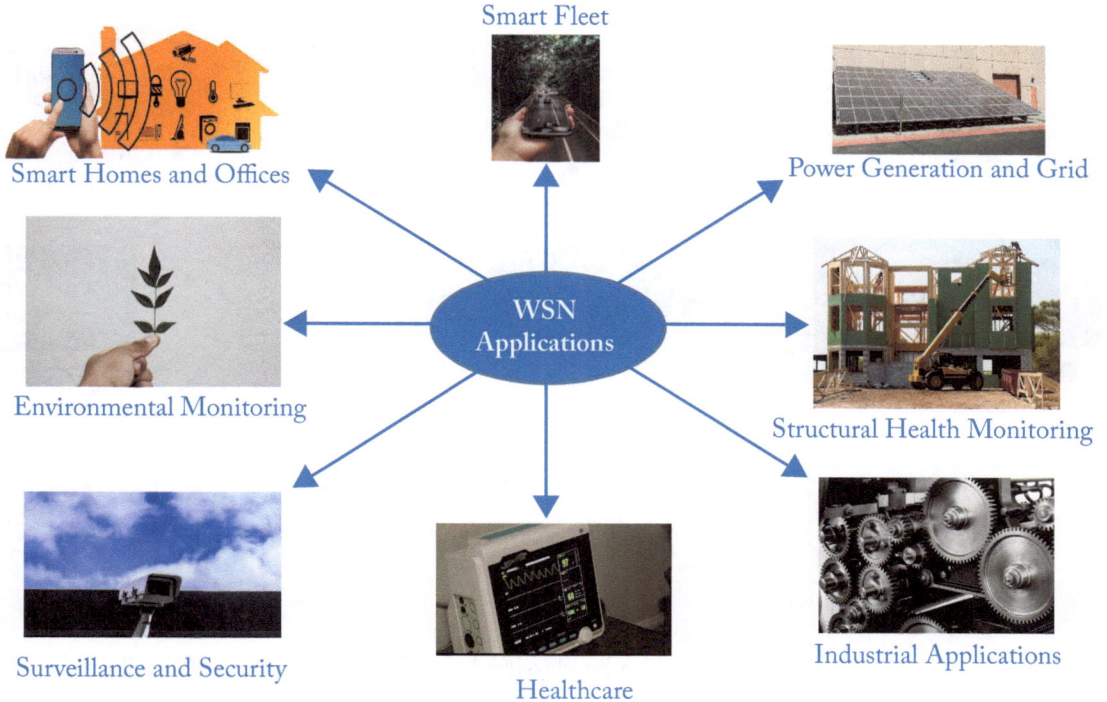

Figure 1.4: Wireless sensor networks (WSNs) have a wide range of applications including smart homes and buildings [40], environmental monitoring [23, 29–34], surveillance and security [6, 25–28, 36–38], healthcare [48–55], smart fleets [40–42, 44–47], and power generation and smart grids [24, 35, 56–59]. More details and further references are provided in Section 1.2.

channels is considered in [20, 85]. A soft-max approximation function, together with a nonlinear average consensus algorithm, is used for noisy max consensus. A more detailed description of max consensus is given in Section 2.2.2.

1.4   NETWORK STRUCTURE ESTIMATION

In this book, algorithms for estimating the structure of WSNs are presented. The term structure refers to how the sensor nodes are connected, how large the coverage area is and where the locations of sensor nodes are. Learning the structure of the network is important in many applications as briefly concluded in Figure 1.5 [59, 86–88]. For example, the professional relations between people can be inferred from the graph structure and connectivity information in a social network [89]. In distributed diffusion adaptation and consensus algorithms, the convergence speed depends on the connectivity of the network [20, 75]; network structure information can be used to derive optimal parameters and accelerate convergence [69]. In some applications, the

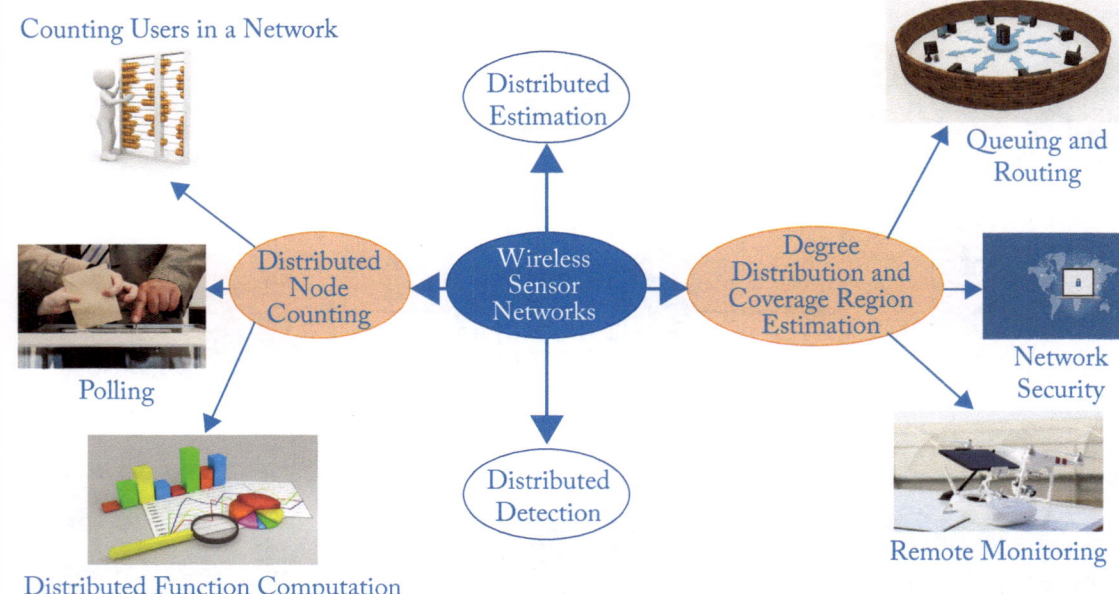

Figure 1.5: Wireless sensor networks can be classified based on applications: Distributed esti-mation, distributed detection, distributed node counting, and degree distribution and coverage region estimation. In this book, we focus on distributed node counting and degree distribution and coverage region estimation. Applications of node counting [27, 28, 92] include counting the number of users in a network, real-time polling, and distributed function computation. De-gree distribution and coverage area can be used to determine the shape of the network as well as the region of influence, with applications [27, 28, 46, 59, 86–88, 92] in queuing and routing, network security, and remote monitoring. More details and further references are provided in Section 1.4.

structure and properties of a small network or subnetwork are used for inferring the structure of a large and complete network [87]. Moreover, network structure and network size information are helpful for locating a service center in a network [90]. It is stated in [91] that the knowledge of the area and the total number of nodes in the network can be used to design optimal con-nections between sensor nodes. In [86] it is shown that energy-efficient scheduling in a WSN depends on the structure and coverage area of the network. Knowledge of the network center and region can be used to place anchor nodes with known locations and then estimate the loca-tions of other sensor nodes [46]. An example of using the known source at the center to localize sensor nodes is given in Figure 1.6. The coverage area and border information of the network are also used in border surveillance applications such as those in [27, 28, 92], where the number of sensors required depends on the size of the coverage region.

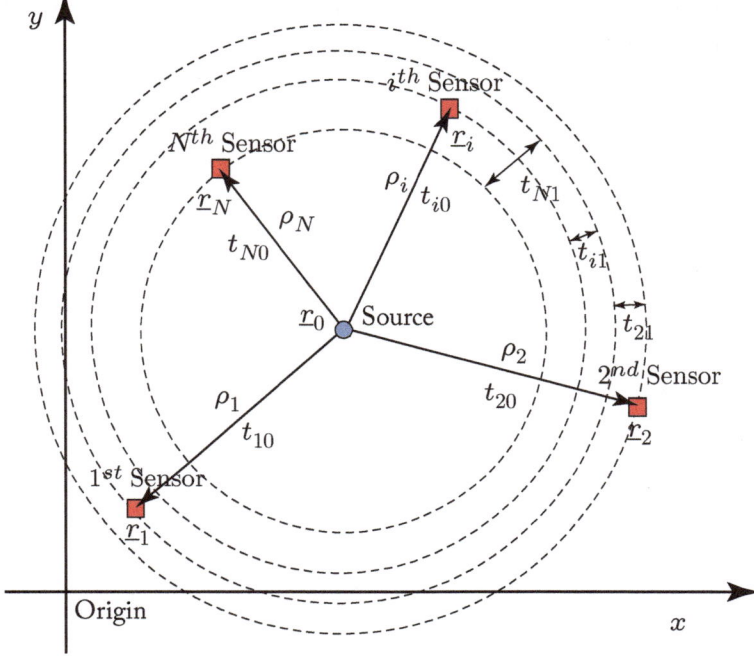

Figure 1.6: WSNs and localization of sensor nodes [46].

A detailed review on network structure estimation algorithms is given in Section 2.3. Many of the existing studies are centralized or depend on the parameters of the network structure. In a fully distributed WSN without a fusion center, it is often hard to learn the structure of the whole network, especially when the network size is large [87].

1.5 ORGANIZATION OF THE BOOK

The main contribution of this book is to review consensus estimation literature and describe three consensus-based distributed network structure estimation algorithms, namely: (a) distributed node counting; (b) degree distribution estimation; and (c) network center and radius estimation. The rest of this book is organized as follows. In Chapter 2, concepts and background knowledge associated with distributed consensus and network structure estimation are reviewed. Chapter 3 describes a distributed node counting algorithm for estimating the system size of a distributed WSN. A distributed network degree distribution estimation algorithm is introduced in Chapter 4. In Chapter 5, a distributed network center and coverage area estimation algorithm is discussed in detail. A discussion on using the introduced algorithms to learn the global data information in distributed networks is also given in Chapter 5. Simulations validating analytical results are also presented. Chapter 6 summarizes the topics covered and provides several references for further reading.

CHAPTER 2

Review of Consensus and Network Structure Estimation

2.1 GRAPH REPRESENTATION OF DISTRIBUTED WSNS

A distributed wireless sensor network is usually modeled as a connected undirected graph, $\mathbb{G} = (\mathbb{N}, \mathbb{E})$. We use $\mathbb{N} = \{1, \ldots, N\}$ to represent the set of nodes, and each sensor is represented by a node on the graph. We use $\mathbb{E}$ to represent the edges between nodes, and any two sensor nodes can communicate with each other if there is an edge between them; we say that they are neighbors. Usually, in a wireless sensor network, two nodes are neighbors if they are physically close to each other. We use $\mathbb{N}_i$ to denote the set of neighbors of node i, that is, $\mathbb{N}_i = \{j | \{i, j\} \in \mathbb{E}\}$.

The full connectivity structure of a wireless sensor network is characterized by the adjacency matrix and the Laplacian matrix. The adjacency matrix $\mathbf{A} = \{a_{ij}\}$ is an $N \times N$ square matrix. The element at row i and column j is $a_{ij} = 1$ if node i and node j are connected, and $a_{ij} = 0$ if there is no edge between the two nodes. A degree matrix, $\mathbf{D} = \text{diag}[d_1, d_2, \ldots, d_N]$, is a diagonal matrix, and its ith diagonal element d_i is the degree of node i (number of neighbors of node i). The Laplacian matrix of a WSN is defined by the adjacency matrix and the degree matrix. The Laplacian matrix $\mathbf{L}$ is defined as $\mathbf{L} = \mathbf{D} - \mathbf{A}$. A simple graph example and their corresponding adjacency matrix, degree matrix, and Laplacian matrix are given in Figure 2.2.

The eigenvalues of the Laplacian matrix are important metrics for characterizing the network structure. For a connected network, the eigenvalues are non negative and the smallest eigenvalue always equals zero, i.e., $\lambda_1(\mathbf{L}) = 0$ and $\lambda_i(\mathbf{L}) > 0, i = 2, \cdots, N$. The eigenvector correspond to $\lambda_1(\mathbf{L}) = 0$ is an all-one vector, that is, $\mathbf{L1} = \mathbf{0}$. The second smallest eigenvalue $\lambda_2(\mathbf{L})$ is an important metric and is called the algebraic connectivity of the graph. $\lambda_2(\mathbf{L})$ is usually used to characterize the density of the network, and for a network with fixed size (number of nodes), a larger $\lambda_2(\mathbf{L})$ indicates a more connected graph. The performance of distributed computation algorithms such as the average consensus algorithm and the diffusion adaptation algorithm often depend on $\lambda_2(\mathbf{L})$ [66]. A study on algebraic connectivity of graphs is given in [93], with several approximations and lower and upper bounds given. For a connected graph, one of the simplest upper bounds on $\lambda_2(\mathbf{L})$ is

$$\lambda_2(\mathbf{L}) \leq \frac{N}{N-1} \min_i d_i, \tag{2.1}$$

where d_i is the degree of node i and N is the total number of nodes in the sensor network. Centralized and distributed methods for estimating $\lambda_2(\mathbf{L})$ are presented in [94, 95]. In Figure 2.1 and Table 2.1, some special graphs (network structure) and their corresponding second largest eigenvalues are presented.

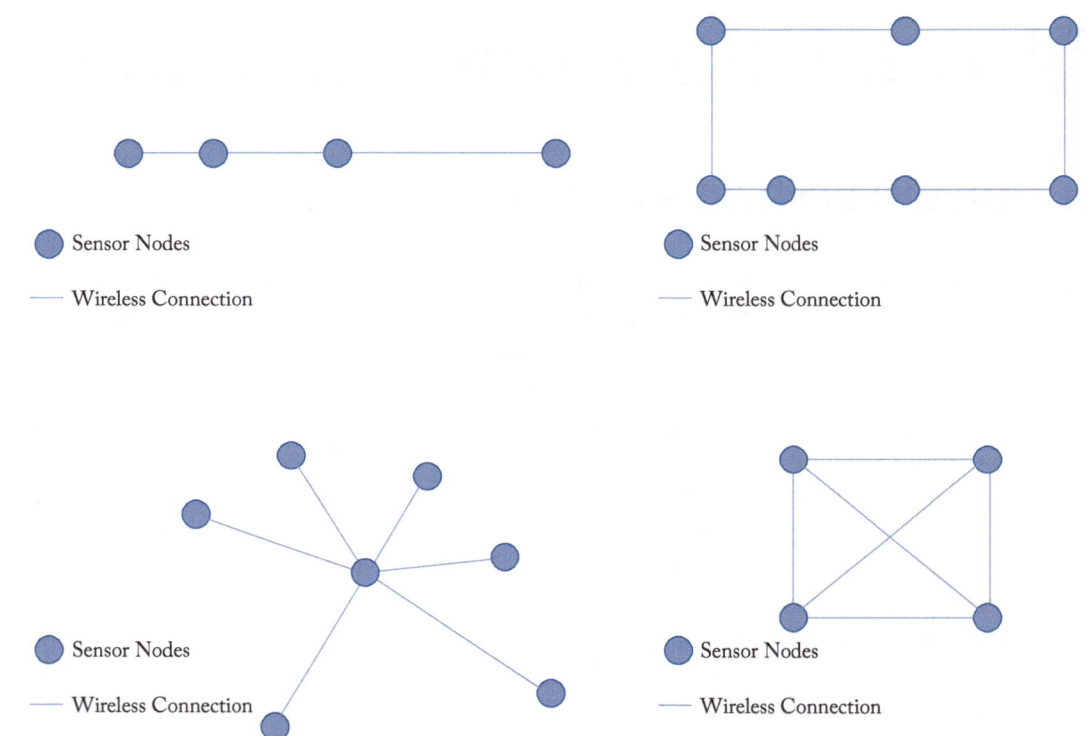

Figure 2.1: Example of some special graphs: path graph (top left), circuit graph (top right), star graph (bottom left), and complete graph (bottom right).

Table 2.1: $\lambda_2(\mathbf{L})$ value for some special graphs

Graph	$\lambda_2(\mathbf{L})$
Path	$2(1 - \cos(\pi/N))$
Circuit	$2(1 - \cos(2\pi/N))$
Star	1
Complete Graph	N
Cube (m – dimensional)	2

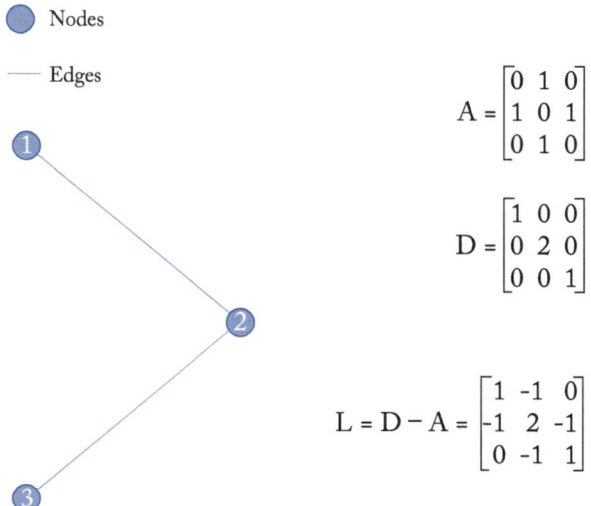

$$A = \begin{bmatrix} 0 & 1 & 0 \\ 1 & 0 & 1 \\ 0 & 1 & 0 \end{bmatrix}$$

$$D = \begin{bmatrix} 1 & 0 & 0 \\ 0 & 2 & 0 \\ 0 & 0 & 1 \end{bmatrix}$$

$$L = D - A = \begin{bmatrix} 1 & -1 & 0 \\ -1 & 2 & -1 \\ 0 & -1 & 1 \end{bmatrix}$$

Figure 2.2: Graph representation of a simple connected undirected network. **A**, **D**, and **L** are the corresponding adjacency matrix, degree matrix, and Laplacian matrix, respectively, as defined in Section 2.1.

2.2   REVIEW OF CONSENSUS ALGORITHMS

Consensus algorithms are well studied [61, 66] and used in distributed and wireless sensor networks for achieving global agreement using only local communications. In this section, we review two of the most popular consensus algorithms: average consensus and max consensus, which will be used in the network structure estimation algorithms introduced in this book.

2.2.1   AVERAGE CONSENSUS

In average consensus problems, it is desired that nodes reach consensus on the global average of all sensed data based on only local communications [66, 96]. Distributed average consensus is used in different areas such as sensor fusion [69, 75, 97, 98], network synchronization [71, 77, 99], distributed detection and estimation [100–102], and distributed function computation [7, 103, 104]. The system model with a brief description of average consensus is given in Figure 2.3.

The basic linear average consensus algorithm is considered in [21, 69, 105]. It is assumed that nodes can communicate only with their neighbors and there is no communication noise between nodes. Assume that the initial state at node i is $x_i(0)$, and $\mathbf{x}(0) = [x_1(0) \cdots x_N(0)]^T$. The iterative updating algorithm for average consensus can be expressed as

$$x_i(t + 1) = \mathbf{W}_{ii}x_i(t) + \sum_{j \in \mathcal{N}_i} \mathbf{W}_{ij}x_j(t), \tag{2.2}$$

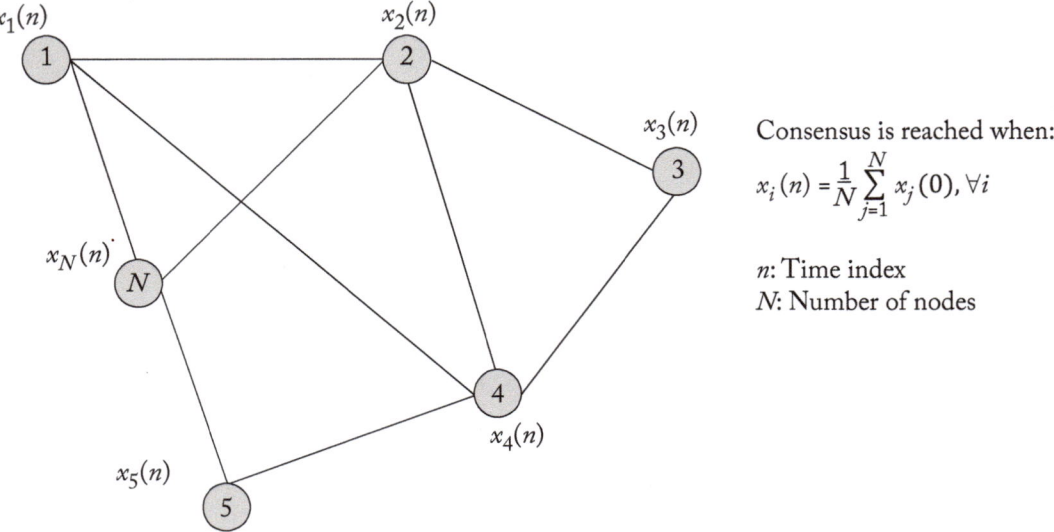

Consensus is reached when:

$$x_i(n) = \frac{1}{N}\sum_{j=1}^{N} x_j(0), \forall i$$

n: Time index
N: Number of nodes

Figure 2.3: Network model for average consensus.

where $i = 1, \cdots, N$ is the node index and $t = 0, 1, 2, \cdots$ is the iteration time index. $\mathbf{W} \in \mathbb{R}_{N \times N}$ is the weight matrix and $\mathbf{W}_{ij}$ is its element in the ith row and jth column. In each iteration, node i updates its state based on its own state at the previous iteration and data from its neighbors, $j \in \mathcal{N}_i$ at the previous iteration.

It can be shown that the convergence of the algorithm is guaranteed if the following conditions are satisfied [69],

$$\mathbf{1}^T\mathbf{W} = \mathbf{1}^T, \ \mathbf{W1} = \mathbf{1}, \tag{2.3}$$

$$\rho\left(\mathbf{W} - \mathbf{1}\mathbf{1}^T\right) < 1, \tag{2.4}$$

where $\rho(\cdot)$ is the spectral radius of a matrix. The convergence speed of linear average consensus algorithm depends on the weight matrix $\mathbf{W}$. A simple way to set the weight matrix is to set the edge weights equal to a constant w, which can be expressed as,

$$\mathbf{W}_{ij} = \begin{cases} w, & \{i, j\} \in \mathbb{E}, \\ 1 - d_i w, & i = j, \\ 0, & \text{otherwise}, \end{cases} \tag{2.5}$$

where the constant $w < 1/d_{\max}$, d_i is the degree of node i and $d_{\max}$ is the maximum degree of all nodes. Algorithms for computing optimal weight matrix for faster convergence are discussed in [67, 69].

In wireless systems, communications between nodes or devices is usually imperfect with noise and interference [64, 78, 106]. Distributed average consensus with communication noise is considered in [64]. It is assumed that communications between nodes is noisy and the network topology changes over time. To compute the average of the initial state $\mathbf{x}(0) = [x_1(0) \cdots x_N(0)]^T$, the average consensus algorithm is used as follows:

$$x_i(t + 1) = [1 - \alpha(t)d_i] x_i(t) + \alpha(t) \sum_{j \in \mathbb{N}_i} [x_j(t) + n_{ij}(t)], \qquad (2.6)$$

where $i = 1, 2, \ldots, N$, and $t = 0, 1, 2, \ldots$, is the time index. The value $x_i(t + 1)$ is the state update of node i at time $t + 1$ and $n_{ij}(t)$ is the noise associated with the reception of $x_j(t)$. $\alpha(t)$ is a positive weight factor, and is a decreasing function of t.

The following assumptions are made in [64]:

Assumptions:

(A1) Mean Connected Graph: Let the Graph Laplacian at time t be $\mathbf{L}(t)$. We assume the Laplacian matrices have mean $\bar{\mathbf{L}} = \mathbf{E}[\mathbf{L}(t)]$, such that $\lambda_i(\bar{\mathbf{L}}) > 0, i = 2, \cdots, N$.

(A2) Independent Noise Sequence: The reception noise, $n_{ij}(t)$, is an independent sequence and the noise is zero mean with bounded variance, that is,

$$\mathbf{E}[n_{ij}(t)] = 0, 1 \le i, j \le N, t \ge 0, \qquad (2.7)$$
$$\mathbf{E}[n_{ij}^2(t)] \le \infty. \qquad (2.8)$$

(A3) Persistence Condition: The positive weight step $\alpha(t)$ is a decreasing function of t and satisfies the conditions:

$$\alpha(t) > 0, \sum_{t=0}^{\infty} \alpha(t) \to \infty, \sum_{t=0}^{\infty} \alpha^2(t) < \infty. \qquad (2.9)$$

It is shown that when assumptions **(A1)**, **(A2)**, and **(A3)** hold, by running the iterative algorithm as in Equation (2.6), the states of nodes converge to a random variable θ, such that

$$\Pr\left[\lim_{t \to \infty} \mathbf{x}(t) = \theta \mathbf{1}\right] = 1. \qquad (2.10)$$

The convergence result, θ, is an unbiased estimator for the average of the initial measurements, $\mathbf{E}[\theta] = \bar{x} = \frac{1}{N} \sum_{i=1}^{N} x_i(0)$, and the mean square error is always bounded, $\mathbf{E}[(\theta - \bar{x})^2] < \infty$.

A distributed average consensus algorithm with bounded transmission power is proposed in [78, 79]. A nonlinear function is used to limit the transmission power of sensors since sensors in the WSN are usually low cost with low power consumption. The update of node i at iteration time $(t + 1)$ is

$$x_i(t + 1) = x_i(t) - \alpha(t) \sum_{j \in \mathbb{N}_i} [h(x_i(t)) - h(x_j(t)) + n_{ij}(t)]. \qquad (2.11)$$

The main difference between Equations (2.11) and (2.6) is the use of a nonlinear function, $h(\cdot)$. The nonlinear function is designed to be monotonically increasing with bounded maximum and minimum value. In [78], sigmoid nonlinear functions are used and it is shown that the consensus result is an unbiased estimator for the global average. Some typical sigmoid functions are shown in Figure 2.4.

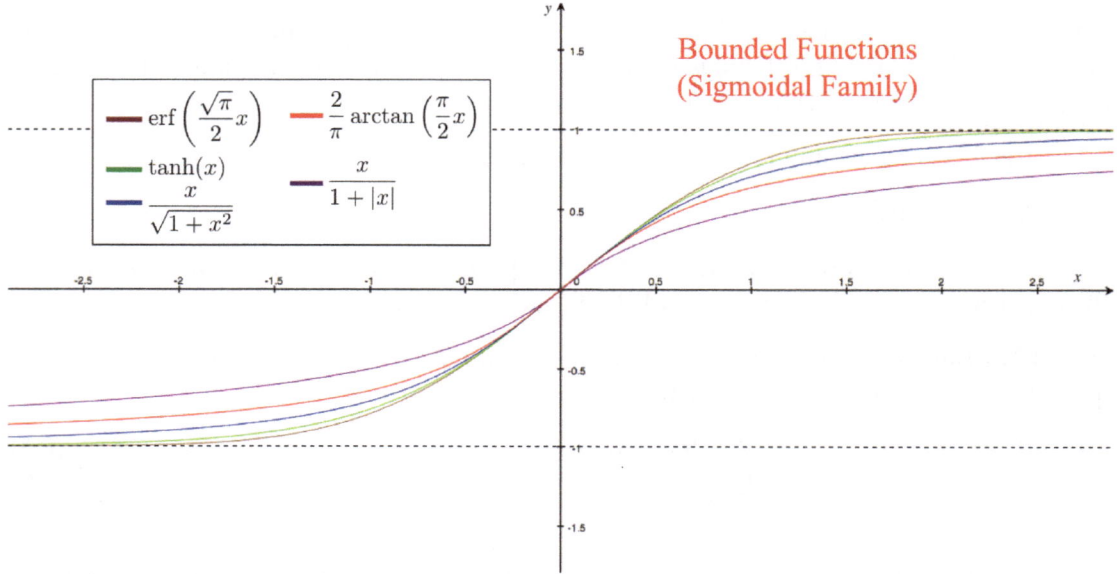

Figure 2.4: Examples of nonlinear sigmoid functions.

2.2.2   MAX CONSENSUS

The problem of estimating the maximum value in a distributed wireless sensor network has been studied extensively. References [66, 80, 82–84] consider max consensus using the max operator. At each time step, every sensor in the network updates its state with the largest measurement it has recovered so far. Let $x_i(0)$ be the initial measurement at node i. The updating rule can be expressed as

$$x_i(t + 1) = \max \left\{ x_i(t), \max_{j \in \mathcal{N}_i} x_j(t) \right\}. \tag{2.12}$$

By running the iterative algorithm in Equation (2.12), nodes in the network converge to the maximum of the initial values in finite iteration time.

In [107], the problem of using consensus to compute general functions in a distributed network is considered, and computing the maximum of the data is one of the functions. A max consensus algorithm based on the weighted power mean algorithm is described. The maximum of $\mathbf{x}(0) = [x_1(0) \cdots x_N(0)]^T$ can be approximated using the weighted power mean function originally proposed by [108]:

$$\max_i x_i(0) = \lim_{r \to \infty} \left(\sum_{i=1}^{N} w_i x_i^r(0) \right)^{\frac{1}{r}} \tag{2.13}$$

where $\sum_{i=1}^{N} w_i = 1$ and r are design parameters in the weighted power mean function. We need to set $r \to \infty$ in the max consensus algorithm. The dynamic behavior of node i is described by u_i, and is given by [107]:

$$u_i = \frac{1}{w_i} x_i^{1-r} \sum_{j \in \mathcal{N}_i} (x_j - x_i), \tag{2.14}$$

where x_i is the state of node i during the continuous max update.

In [20, 81, 85], distributed max consensus is achieved by using a log–sum–exponential approximation and average consensus. A pre-processing exponential function is used for mapping the initial measurements, $y_i(0) := e^{\beta x_i(0)}$, where $\beta > 0$ is a design parameter. Then, the max consensus update at node i at time $(t + 1)$ can be expressed as

$$y_i(t+1) = y_i(t) - \alpha(t) \left[d_i h(y_i(t)) - \sum_{j \in \mathbb{N}_i} h(y_j(t)) + n_i(t) \right], \tag{2.15}$$

where $n_i(t)$ is the noise received at node i at time t, and $h(\cdot)$ is a nonlinear sigmoid function to bound the transmit power. Assuming consensus is reached at time t^*, the estimated maximum can be calculated by applying the following post processing function:

$$\hat{x}_{\max_i}(t^*) = \frac{1}{\beta} \left(\log N + \log y_i(t^*) \right). \tag{2.16}$$

Note that when set $\beta < 0$, the above algorithm can be used to estimate the minimum of the initial measurements. The performance of the algorithm depends on the value of β, and there is a trade off between the convergence speed and the estimation error. A graph example of the max consensus algorithm using Equations (2.15) and (2.16) is given in Figure 2.5.

Other studies that consider the problem of max consensus are given in [7, 82, 83]. The problem of computing functions such as maximum, average, or norm of the data at sensor nodes in centralized wireless sensor network is considered in [7]. To compute the maximum, a maximum approximation function is used in [7], and the max of all sensor data is estimated at the fusion center. In [82], a directed graph/network is considered, and a max-plus algebra is used to analyze the max consensus algorithm in the directed graph. In [83], it is assumed that the network structure is changing over time and the problem of distributed maximum estimation is considered. It is shown in [83] that max consensus can be obtained if strong connectivity of the graph is satisfied.

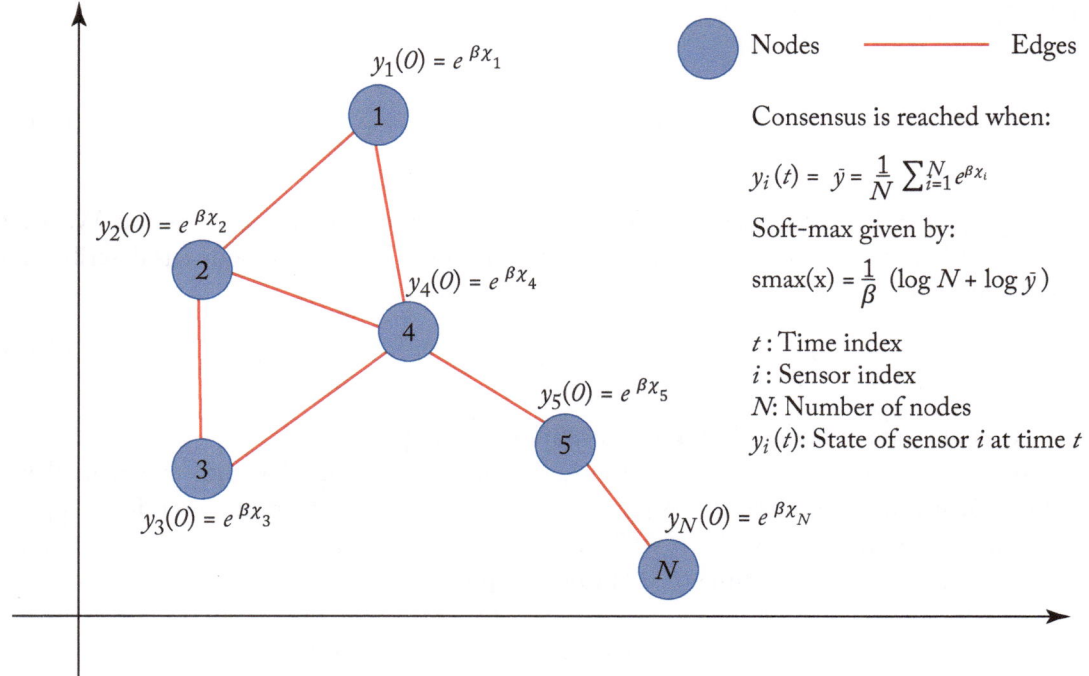

Figure 2.5: Graph example of max consensus using log-sum-exponential approximation as in [20].

2.3 REVIEW OF NETWORK STRUCTURE ESTIMATION

Learning the structure of a wireless sensor network is important and useful in many applications such as accelerating consensus algorithms [67, 69, 78], setting front/back-end and processing operations in networks [109, 110], and incorporating newly added nodes [111]. In this section, existing network structure estimation approaches such as network connectivity state estimation, system size estimation, and network coverage region estimation are described.

2.3.1 NETWORK CONNECTIVITY STATE ESTIMATION

The connectivity state of a wireless sensor network is characterized by the adjacency matrix as described in Section 2.1. In [112], an average consensus-based adjacency matrix estimation algorithm is proposed. It is assumed that all nodes in the network are labeled and have unique IDs. Each node i first detects its direct neighbors to calculate the initial local estimate of the adjacency matrix, $\hat{\mathbf{A}}_i(0)$. Then average consensus is used, and the update of node i at iteration time $(t + 1)$ can be expressed as

$$\hat{\mathbf{A}}_i(t + 1) = \hat{\mathbf{A}}_i(t) + \alpha \sum_{j \in \mathcal{N}_i} \left(\hat{\mathbf{A}}_j(t) - \hat{\mathbf{A}}_i(t) \right), \tag{2.17}$$

where $\alpha > 0$ is the update step size. Assuming consensus is reached at time t^*, an estimate of the adjacency matrix can be obtained by setting the non zero entries of $\hat{\mathbf{A}}_i(t^*)$ to 1 and 0 otherwise. Note that fading and communication noise are also considered in [112]. Procedures to pre-map and post process the results are also described. However, the algorithm is not fully distributed in the sense that the nodes need to be labeled.

In fully distributed wireless sensor networks, where sensor nodes are not labeled, estimating the network adjacency matrix is challenging. Recall that the second smallest eigenvalue of the Laplacian matrix, $\lambda_2(\mathbf{L})$, usually characterizes the density and connectivity of a network [93, 113]. Therefore, the problem of estimating the algebraic connectivity of the graph, $\lambda_2(\mathbf{L})$, attracts many researchers [94, 95, 114]. The decentralized algebraic connectivity estimation is inspired by the power iteration algorithm, in other words, the eigenvector corresponding to algebraic connectivity, $\lambda_2(\mathbf{L})$, can be calculated using iterative algorithms. Let $\mathbf{x} = (x_1 \cdots x_N)^{\mathrm{T}}$ be the states of nodes. Consider the iteration algorithm:

$$\dot{x}_i = -k_1 z_{i,1} - k_2 \sum_{j \in \mathcal{N}_i} (x_i - x_j) - k_3(z_{i,2} - 1)x_i, \tag{2.18}$$

where $\dot{x}_i$ is the dynamic of the state of node i. k_1, k_2 and k_3 are design parameters, $z_{i,1}$ is the average estimate of all x_i and $z_{i,2}$ is the average estimation of all x_i^2. Note that average consensus can be used to calculate $z_{i,1}$ and $z_{i,2}$. When the iteration time is large, the estimate of $\lambda_2(\mathbf{L})$, denoted as $\hat{\lambda}_2(\mathbf{L})$, can be calculated as:

$$\hat{\lambda}_2(\mathbf{L}) = \frac{k_3}{k_2}(1 - z_{i,2}). \tag{2.19}$$

Note that estimating $\hat{\lambda}_2(\mathbf{L})$ and the corresponding eigenvector are widely used in network clustering applications [94, 95, 115, 116].

2.3.2 SYSTEM SIZE ESTIMATION

System size estimation or estimating the number of active sensor nodes in a wireless sensor or distributed network is essential in several applications. For example, it is mentioned in [111] that the network size information can be used to maintain a network and incorporate newly encountered sensor nodes. In some distributed function computation algorithms such as aggregation of local values [103] and max/min/average estimation [20, 85], the system size is also needed. It is shown in [117–119] that in a cognitive radio system, the number of primary and secondary users affects the performance of the system. Estimating the system size is a challenging problem in a decentralized WSN, where sensors only have local information.

One of the most popular methods for system size estimation is random sampling [111, 120–125]. It is usually assumed that nodes are labeled. In [120, 121], a random walk method is used to achieve statistical properties similar to independent sampling among the nodes in the network. The selected sender sends out information to one of its neighbors, and nodes in the network pass the information. The information propagates via a random walk. Based on the return time, the number of nodes in the network can be estimated.

In other approaches, nodes estimate the system size without labeling assumptions. In [103], probabilistic counting algorithms, originally used for estimating the number of distinct elements in a database, are introduced. It is shown that the probabilistic counting approach can be used to estimate the system size in a distributed WSN. The node counting problem is solved by generating a random bit sequence followed by in-network computation to estimate the total number of distinct initial bit sequences. A comparative study of several system size estimation algorithms is given in [127]. Several system size estimation algorithms that rely on the knowledge of graph structure are given in [128] and [129].

Some recent studies consider using average and max consensus to estimate the number of nodes in the network. An average consensus-based approach is introduced in [130]. The initial values at nodes are generated from a Bernoulli distribution and the average consensus result is used. The smallest integer that divides the consensus result is defined to be the estimate of the system size in one consensus run. Then, by running the algorithm multiple times, the system size N can be determined from the least common multiple of estimates obtained from different consensus runs. Two other algorithms based on average and max consensus are proposed in [131–133], where by generating specific random variables at nodes, the number of nodes can be inferred from the average consensus results or max consensus results. In [131, 133], a uniform + maximum + ML algorithm is described. The initial values are uniformly distributed and the system size can be calculated from the max consensus results. In the Gaussian + average + ML algorithm in [131], the initial values at nodes are randomly generated from a Gaussian distribution and linear average consensus is used. Note that all above mentioned algorithms assume the communications between nodes are perfect without noise. In [134, 135], it is assumed that there is a leader in the network or max consensus is first used to decide a leader in the graph, and the initial states of all nodes except the leader are set to zero, while the initial state of the leader is set to be one. Then average consensus is used for node counting.

2.3.3 NETWORK COVERAGE REGION ESTIMATION

The coverage region or the area of the network is also an important metric to characterize the structure of the network [136–138]. Research considers the network coverage region estimation can be found in [126, 137–140]. It is assumed in [126, 139] that each node can detect its own sensing region and the total sensing area for all nodes can be computed. The sensing region for each node can be represented using different ways, for example, it is assumed in [126] that every node in the network covers a circular region with same radius. A "digital circle" method is used to represent the real circle sensing region for each node and an algorithm for estimating the union of all "digital circle" area is proposed to compute the whole network coverage region. An example of the coverage region of network defined by total circle sensing area [126] is shown in Figure 2.6. In [137, 138], the network coverage region is defined to be the smallest ellipse that contains all sensor nodes. Centralized algorithms for estimating the ellipsoidal or spherical coverage region are described. In [140, Chapter 8], the problem of estimating the inner and outer extremal volume ellipsoid of a point set (can be treated as nodes in the network) is formulated as optimization algorithms. The problem of estimating the convex hull of a point set is also discussed.

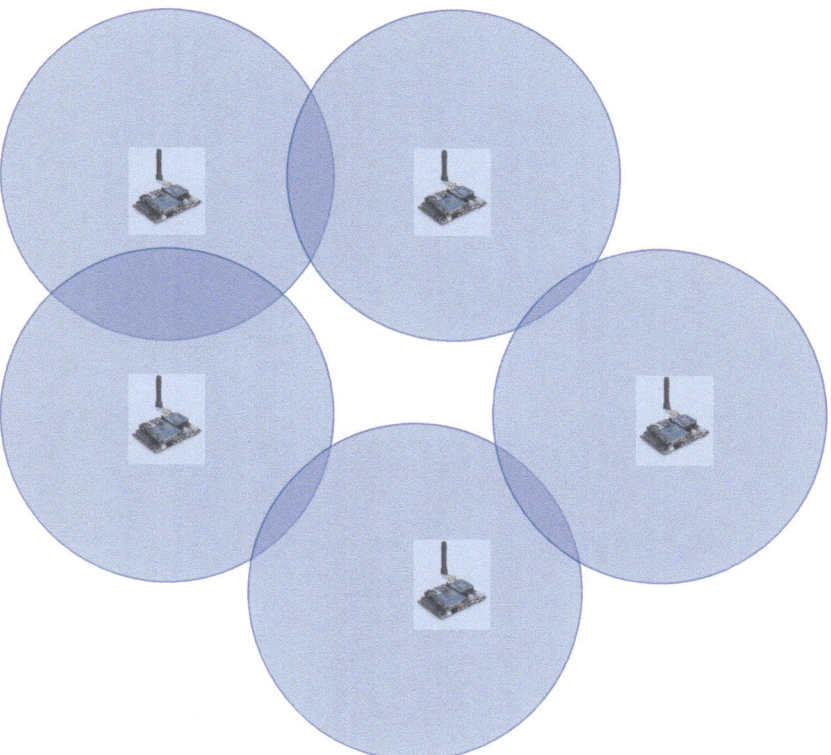

Figure 2.6: Area coverage by "digital circles" proposed in [126]. A "digital circle" is used to represent the real circle to make the computation simple and accurate.

CHAPTER 3

Distributed Node Counting in WSNs

In this chapter, a distributed connected network with N nodes is considered. A distributed algorithm for counting the total number of nodes in a wireless sensor network with noisy communications is introduced. Note that part of the content in this chapter is based on our published work [141–143].

3.1 SYSTEM MODEL

The distributed wireless sensor network is modeled as an undirected connected graph with N nodes as described in Section 2.1. We assume that the nodes in the network have limited memory and no knowledge about the structure of the graph, and also that nodes can communicate only with their neighbors. We also assume analog transmissions between nodes [64, 69, 78] and presence of communication noise, which is independent and identically distributed (i.i.d) with zero mean and variance σ_n^2.

3.2 DISTRIBUTED NODE COUNTING BASED ON L_2 NORM ESTIMATION

Average consensus in which the states of the nodes converge to the sample mean of the initial states was reviewed in Section 2.2.1. Assume $\mathbf{x} = [x_1(0) \cdots x_N(0)]^T$ contains the initial states of nodes. The key equation that relates the network size N to average consensus is to observe that

$$N = \frac{||\mathbf{x}||_2^2/N}{||\mathbf{x}||_2^2/N^2}. \tag{3.1}$$

The value of the numerator and denominator in (3.1) are scaled versions of the L_2 norm of $\mathbf{x}$. As will be seen later in this section, average consensus with designed random initial values at nodes can be used to estimate the value of the numerator and denominator in (3.1), and an estimate of the number of nodes in the network can be obtained. The node counting algorithm can be described in three phases: an estimate of the value of the denominator and numerator in (3.1) can be obtained using the average consensus algorithm in phase I and phase II, respectively, and $\hat{N}$ is calculated in phase III by using the consensus results of phases I and II to compute the ratio in (3.1). In what follows, details of the three phases of the algorithm are described.

3.2.1   PHASE I: L_2 NORM ESTIMATION

In Phase I of the node counting algorithm, an estimate of the denominator in Equation (3.1) is calculated based on L_2 norm estimation using average consensus algorithm. Assume the initial values of nodes are $\mathbf{x} = [x_1 \cdots x_i \cdots x_N]$, where x_i is the initial value at node i. Then using the initial value, each node in the network generates a $K-$dimensional initial state vector. The initial state vector at node i is denoted as $\mathbf{y}_i(0) = [y_i^{(1)}(0) \cdots y_i^{(K)}(0)]$, where $y_i^{(k)}(0) = r_i^{(k)} x_i$, $1 \leq k \leq K$, and $r_i^{(k)}$ are i.i.d. random variables with zero mean and variance one generated at nodes.

By running the average consensus algorithm, each node updates the kth element in the state vector of node i at time $t + 1$ with

$$y_i^{(k)}(t+1) = [1 - \alpha(t)d_i]\, y_i^{(k)}(t) + \alpha(t) \sum_{j \in \mathbb{N}_i} \left[y_{ij}^{(k)}(t) + n_{ij}^{(k)}(t) \right], \qquad (3.2)$$

where $n_{ij}^{(k)}(t)$ is the noise associated with the reception of $y_{ij}^{(k)}(t)$, and $\alpha(t)$ satisfies the persistence condition in Equation (2.9). When t is large, the kth element of node i converges to a noisy version of the average, $\frac{1}{N} \sum_{i=1}^{N} r_i^{(k)} x_i$.

A postprocessing function $f(\cdot)$ is applied at each node by squaring each element in the state vector and taking the average of the result. It can be shown that the postprocessed result is an unbiased estimator for the denominator in Equation (3.1). Assume that the consensus stops at iteration time t^*. The postprocessed result for node i can be expressed as,

$$f\left(\mathbf{y}_i(t^*)\right) = \frac{1}{K}||\mathbf{y}_i(t^*)||^2 = \frac{1}{K} \sum_{k=1}^{K} \left(y_i^{(k)}(t^*)\right)^2 \approx \frac{1}{N^2}||\mathbf{x}||_2^2 \qquad (3.3)$$

where $\mathbf{y}_i(t) = [y_i^{(1)}(t) \cdots y_i^{(K)}(t)]$ is the state vector of node i at time t.

3.2.2   PHASE II: L_2 NORM CONSENSUS

In Phase II of the node-counting algorithm, the numerator in Equation (3.1) is calculated based on the definition of L_2 norm using the average consensus algorithm. Node i sets its initial state value to $z_i(0) = x_i^2$. Nodes in the network apply average consensus:

$$z_i(t+1) = [1 - \alpha(t)d_i]\, z_i(t) + \alpha(t) \sum_{j \in \mathbb{N}_i} \left[z_{ij}(t) + n_{ij}(t) \right]. \qquad (3.4)$$

Assume that the iterative algorithm stops at t^*. We have,

$$z_i(t^*) \approx \frac{1}{N}||\mathbf{x}||_2^2. \qquad (3.5)$$

3.2.3   PHASE III: NODE COUNTING

By comparing the results from Equations (3.3) and (3.5), the estimate of the number of nodes in the network, $\hat{N}_i(t^*)$, at node i at time t^* can be obtained as

$$\hat{N}_i(t^*) = \frac{z_i(t^*)}{f(y_i(t^*))}. \tag{3.6}$$

Note that Phase I and Phase II can be done at the same time by using a $K + 1$ state vector containing both the $K \times 1$ process $\mathbf{y}_i(t)$ and the scalar $z_i(t)$. An overview of the described node-counting algorithm is given in Figure 3.1. From (3.6), the algorithm works for any $\mathbf{x} \neq \mathbf{0}$.

In the algorithm, x_i can be sensor measurements or values designed for improved performance. Random variables $r_i^{(k)}$ are i.i.d. and generated at nodes with mean 0 and variance 1. Two simple ways to choose $r_i^{(k)}$ are: (i) Normally distributed, $r_i^{(k)} \sim \mathcal{N}(0, 1)$; and (ii) Bernoulli distributed with ± 1, that is, $\Pr[r_i^{(k)} = 1] = \Pr[r_i^{(k)} = -1] = 1/2$ [141].

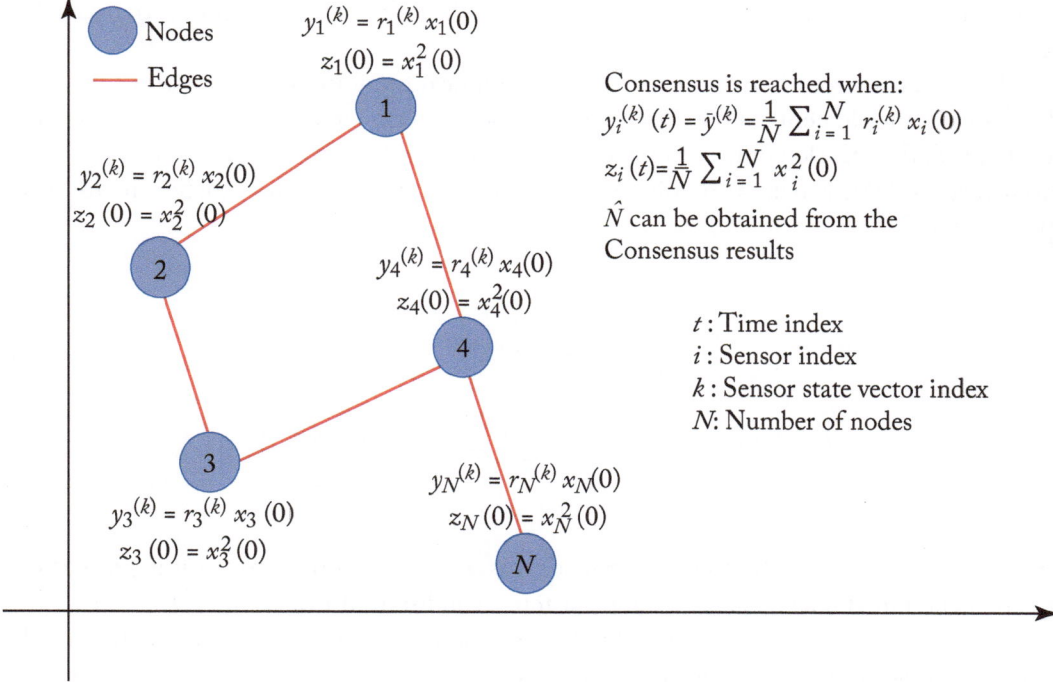

Figure 3.1: An overview of the distributed node counting algorithm.

3.3   PERFORMANCE ANALYSIS

In this section, different sources of error in the algorithm are explicitly discussed. There are three different sources of error in the proposed node counting algorithm: bias caused by lack of convergence, error caused by communication noise, and L_2 norm estimation error caused by the norm approximation function.

Since average consensus is used, the consensus is reached only when the iteration time $t \to \infty$ [64]. The transient of the bias (bias between the state of nodes and the desired consensus result) in the algorithm is caused by lack of convergence in finite iteration time. The nodes ideally converge to the average state vector $\bar{\mathbf{y}} = [\bar{y}^{(1)} \cdots \bar{y}^{(k)} \cdots \bar{y}^{(K)}]^T$, where $\bar{y}^{(k)} = \frac{1}{N} \sum_{i=1}^{N} r_i^{(k)} x_i$. Let $\mathbf{y}^{(k)}(t) = [y_1^{(k)}(t) \ y_2^{(k)}(t) \cdots y_N^{(k)}(t)]^T$ contain the kth element in the state vector from all N nodes at time t. The convergence rate of the mean of, $\mathbf{y}^{(k)}(t)$, is quantified by [64, Eq. (61)]:

$$\left\| \mathbf{E}\left[\mathbf{y}^{(k)}(t)\right] - \bar{y}^{(k)}\mathbf{1} \right\|_2 \leq \left(e^{-\lambda_2(\mathbf{L})\sum_{\tau=0}^{t} \alpha(\tau)}\right) ||\mathbf{y}^{(k)}(0) - \bar{y}^{(k)}\mathbf{1}||_2. \qquad (3.7)$$

It is clear from Equation (3.7) that the convergence is fast if the algebraic connectivity $\lambda_2(\mathbf{L})$ is large.

The second source of error is caused by communication noise. Even though $\mathbf{E}[y^{(k)}(t)] \to \bar{y}^{(k)}$ as $t \to \infty$, the elements in the state vectors do not converge to the true average of the initial states. Instead, the kth elements of the state vectors for all nodes converge *almost surely* to a finite random variable $\theta^{(k)}$, which is an unbiased estimator of the average and satisfies the following properties [64, 78]:

$$\mathbf{E}[\theta^{(k)}] = \frac{1}{N} \sum_{i=1}^{N} r_i^{(k)} x_i = \bar{y}^{(k)}, \qquad (3.8)$$

$$\zeta^{(k)} = \mathbf{E}\left[\left(\theta^{(k)} - \bar{y}^{(k)}\right)^2\right] \leq \frac{(N-1)\sigma_n^2 \sum_{t=0}^{\infty} \alpha^2(t)}{N}, \qquad (3.9)$$

where σ_n^2 is the communication noise variance. It is seen in (3.9) that the error is proportional to σ_n^2 and is bounded if $\alpha(t)$ satisfies the persistence condition in Equation (2.9).

The third source of error is caused by the fact that the L_2 approximation function used in Section 3.2.1 is accurate only when K is large. The consensus result is an unbiased estimator for the scaled L_2 norm of the initial measurements as in Equation (3.3). However the estimation variance depends on the design parameters K, $r_i^{(k)}$ and x_i. Let

$$Y = \frac{1}{K} \sum_{k=1}^{K} \left(\frac{1}{N} \sum_{i=1}^{N} r_i^{(k)} x_i\right)^2. \qquad (3.10)$$

$Z^2 = \left(\frac{1}{N} \sum_{i=1}^{N} r_i^{(k)} x_i \right)^2$ can be proved to be an unbiased estimator of $||\mathbf{x}||_2^2$ [141]. The variance of Z^2 can be calculated as:

$$\text{Var}[Z^2] = \mathbf{E}[Z^4] - \left(\mathbf{E}[Z^2] \right)^2 \tag{3.11}$$

$$= \frac{\left(\mathbf{E}\left[\left(r_i^{(k)} \right)^4 \right] - 1 \right)}{N^4} \sum_i x_i^4 + \frac{4}{N^4} \sum_{i<j} x_i^2 x_j^2. \tag{3.12}$$

Then, the variance of Y can be expressed as:

$$\sigma_Y^2 = \text{Var}[Y] = \frac{1}{K} \text{Var}[Z^2]. \tag{3.13}$$

Equations (3.12) and (3.13) show that the variance will be small when K is chosen to be large, and the variance is also related to $r_i^{(k)}$ and $x_i(0)$. Therefore, there is a trade off between the accuracy of the algorithm and the storage at sensor nodes: A more accurate estimate of number of nodes can be obtained if K is large, but the nodes need to keep a larger state vector and therefore increase the required storage at nodes. More details on performance analysis can be found in [141, 142].

3.4   SIMULATION RESULTS

In this section, simulation results for the node-counting algorithm are given. We first consider a connected network with $N = 75$ as shown in Figure 3.2. In Figures 3.3–3.5, we set $K = 1000$, noise variance $\sigma_n^2 = 1$, and $\alpha(t) = 0.1/(t + 1)$. In Figure 3.3, the initial values of x_i are fixed and generated from a Gaussian distribution with zero mean and variance 25, and $r_i^{(k)}$ are Bernoulli distributed with ± 1 with the same probability so that signal-to-noise ratio SNR $= 13.98$ dB (SNR is defined to be the ratio of power of $y_i^{(k)}$ and power of noise). In Figure 3.4, the initial values are set to be fixed at $x_i = 5$ and $r_i^{(k)}$ are Bernoulli distributed with ± 1, and SNR $= 13.98$ dB. In Figure 3.5, the initial values are set to be fixed $x_i = 5$ and $r_i^{(k)}$ are chosen as $r_i^{(k)} \sim \mathcal{N}(0, 1)$, and SNR $= 13.98$ dB. From Figures 3.3–3.5, we see that the number of nodes can be estimated using the proposed node counting algorithm in the presence of communication noise.

In Figure 3.6, the mean square error for $\hat{N}(t)$, denoted as $\mathbf{E}\left[\left(\hat{N}(t) - N \right)^2 \right]$, is plotted.

The graph is the same as in Figure 3.2 with $N = 75$, and x_i and $r_i^{(k)}$ are chosen as shown in the Figure 3.6. We assume noisy communication with $\sigma_n^2 = 1$ and $K = 1000$. From Figure 3.6, we can see that in the presence of communication noise, larger x_i values result in better performance since the signal to noise ratio is larger. We can also see from Figure 3.6 that for the same x_i value, mean square error (MSE) value are almost the same for different $r_i^{(k)}$ (Bernoulli $r_i^{(k)}$ performs a little bit better, see [141] for detailed explanation).

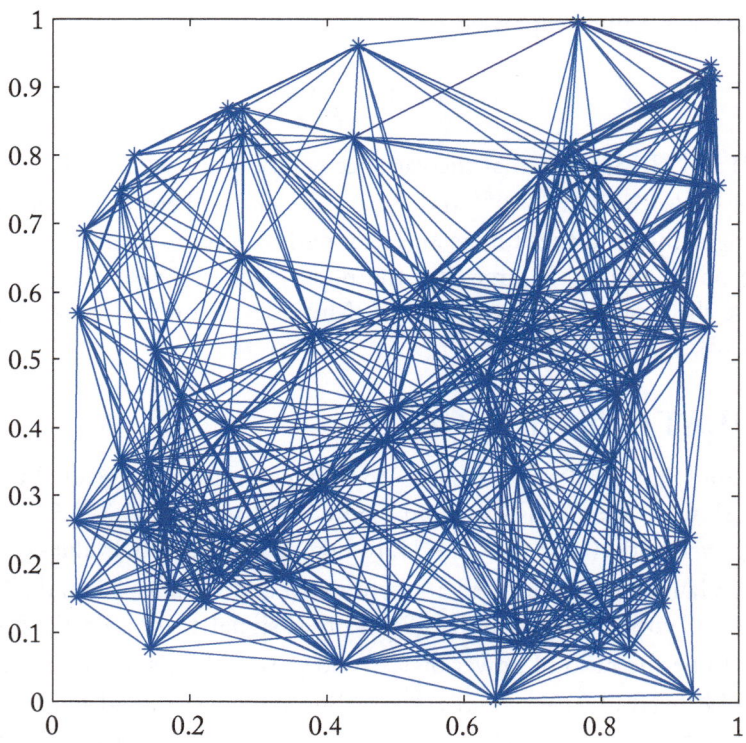

Figure 3.2: Graph representation of the sensor network in simulation, $N = 75$.

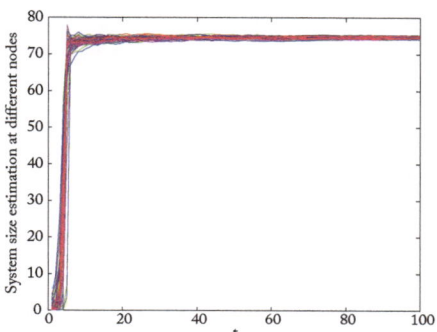

Figure 3.3: Node-counting convergence results vs. number of iterations, t. $x_i(0) \sim \mathcal{N}(0, 25)$, $\sigma_n^2 = 1$, $r_i^{(k)}$ Bernoulli distributed with ± 1, $\alpha(t) = 0.1/(t + 1)$ and $K = 1000$.

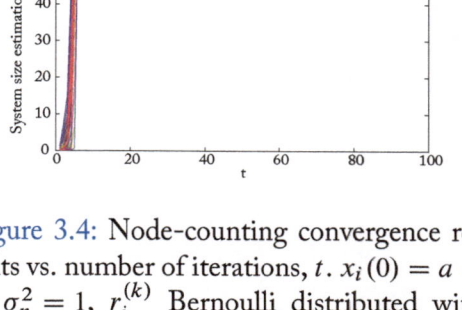

Figure 3.4: Node-counting convergence results vs. number of iterations, t. $x_i(0) = a = 5$, $\sigma_n^2 = 1$, $r_i^{(k)}$ Bernoulli distributed with ± 1, $\alpha(t) = 0.1/(t + 1)$ and $K = 1000$.

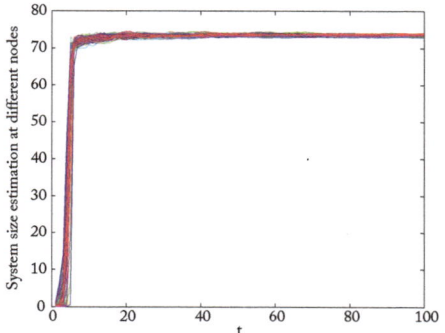

Figure 3.5: Entries of node-counting convergence results vs. number of iterations, t. $x_i(0) = a = 5$, $\sigma_n^2 = 1$ and $r_i^{(k)} \sim \mathcal{N}(0, 1)$, $\alpha(t) = 0.1/(t + 1)$ and $K = 1000$.

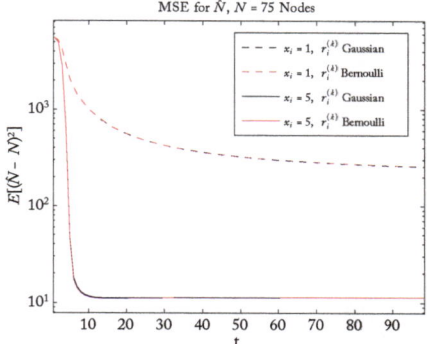

Figure 3.6: MSE vs. t, noise variance $\sigma_n^2 = 1$, and $K = 1000$.

CHAPTER 4

Noncentralized Estimation of Degree Distribution

In this chapter, a distributed network degree distribution estimation (DNDD) algorithm for estimating the network degree distribution and degree matrix of a WSN is introduced. Degree distribution and degree matrix are important metrics to characterize the structure of networks and are used in several applications [144–146]. For example. the knowledge of degree distribution and degree matrix can be used to infer the properties of distributed networks such as minimum/maximum node degree, k−connectivity, and edge connectivity [144, 145]. It is shown in [146] that different network topologies have different network degree distributions. For example, random graphs usually have Poisson degree distribution, whereas in real-world networks, the degree distribution usually follows power-law degree distribution [146]. Therefore the topology of the network such as random topology and tree topology can be inferred from the degree distribution. Note that part of the content in this chapter is presented in [147].

4.1   SYSTEM MODEL

Consider a connected wireless sensor network with no fusion center. The network is modeled as an undirected graph as in Section 2.1. We assume that two nodes can communicate with each other if and only if they are neighbors. We also assume that communication channels between nodes are noisy.

Define n_k to be the number of nodes with degree k, and N the total number of nodes in the network. The degree distribution, $p^{(k)}$, of a network is defined to be $p^{(k)} = \frac{n_k}{N}, k = 1, 2, \cdots, d_{\max}$, where $d_{\max}$ is the maximum degree.

An example of a simple distributed WSN and its degree distribution is given in Figure 4.1. In Figure 4.1, there is one node with degree 1 (the right most node), two nodes with degree 2 (the two nodes on the left), and one with degree 3 (the node in the center). Therefore the degree distribution is as shown in the figure, where the probability that a node has degree equals to k is denoted as $\Pr(d = k)$.

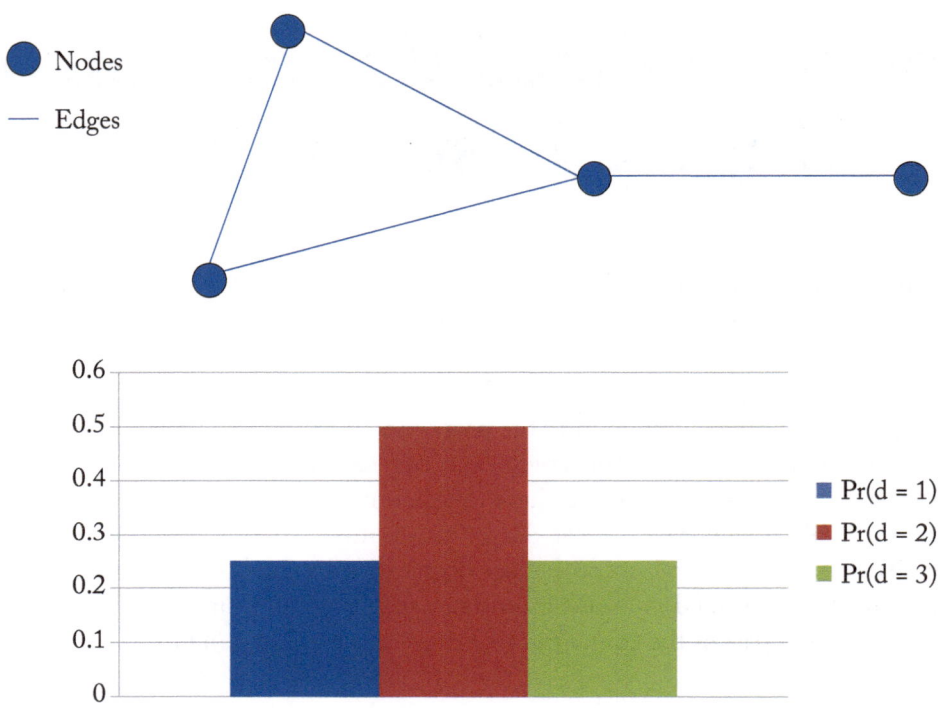

Figure 4.1: A simple WSN and its degree distribution.

4.2   CONSENSUS-BASED DEGREE DISTRIBUTION ESTIMATION

The proposed algorithm is based on the fact that degrees are discrete values and probability mass functions can be estimated with average consensus. In the following, three steps of the proposed algorithm are provided, namely: initial measurement vectors are generated at nodes in step I; average consensus is performed in step II to let nodes reach consensus on the same state vector; and the degree distribution is calculated in step III by postprocessing the convergence result.

4.2.1   STEP I: GENERATE INITIAL VALUES

In step I of the algorithm, each node in the network generates an initial measurement vector with length K, $\mathbf{x}_i(0) = \left[\mathbf{x}_i^{(1)}(0) \quad \mathbf{x}_i^{(2)}(0) \quad \cdots \mathbf{x}_i^{(K)}(0) \right] \in \mathcal{R}^K$. $K > d_{\max}$, where $d_{\max}$ is the maximum degree. Note that max consensus introduced in Section 2.2.2 can be used to estimate $d_{\max}$

in WSNs. Assume that node i only knows its own degree d_i, the kth element of the initial measurement vector at node i, denoted as $\mathbf{x}_i^{(k)}(0)$, which can be expressed as:

$$\mathbf{x}_i^{(k)}(0) = \begin{cases} 1, & \text{if } d_i = k \\ 0, & \text{otherwise,} \end{cases} \tag{4.1}$$

where $k = 1, 2, \cdots, K$. Note that by setting the initial vector as in Equation (4.1), there is one and only one nonzero element in the state vector for each node. The initial vector is generated in a fully distributed way in Equation (4.1) in the sense that each node i only needs to know its own degree d_i and no information about other nodes is needed.

4.2.2   STEP II: AVERAGE CONSENSUS

In step II of the degree distribution estimation algorithm, the average consensus algorithm described in Section 2.2.1 is used for computing the elements in the degree distribution probability mass function. Each node i sets the initial state vector $\mathbf{x}_i(0)$ as in Equation (4.1). Nodes in the network run the average consensus algorithm on each element in the state vectors, and the kth element in the state vector of node i at time $t + 1$ can be expressed as,

$$\mathbf{x}_i^{(k)}(t+1) = [1 - \alpha(t)d_i]\,\mathbf{x}_i^{(k)}(t) + \alpha(t) \sum_{j \in \mathbb{N}_i} \left[\mathbf{x}_j^{(k)}(t) + v_{ij}^{(k)}(t) \right], \tag{4.2}$$

where $v_{ij}^{(k)}(t)$ is the noise associated with the reception of $\mathbf{x}_j^{(k)}(t)$ at node i.

It is shown [64] that by running the update as in Equation (4.2), the state vectors for all nodes converge to a noisy version of the average of the initial measurement vectors. Assume that the iteration stops at time t^*. We have:

$$\mathbf{E}\left[\mathbf{x}_i^{(k)}(t^*)\right] = \frac{1}{N} \sum_{j=1}^{N} \mathbf{x}_j^{(k)}(0) = \left(\frac{n_k}{N}\right), \tag{4.3}$$

where n_k is the number of nodes with degree k.

Equation (4.3) shows that the convergence result for the kth element in the state vector is an unbiased estimator for the kth element in the degree distribution $p^{(k)}$. It is also proved in [64] and [78] that the MSE of the algorithm is bounded and will be small when the noise variance σ_n^2 is small.

Assuming consensus is perfectly reached and there is no communication noise, then the convergence result will be an accurate estimate of the degree distribution.

$$\mathbf{x}_i^{(k)}(t^*) := \hat{p}_i^{(k)}(t^*) = \frac{n_k}{N} = p_i^{(k)}(t^*). \tag{4.4}$$

However, consensus in wireless sensor networks always suffers from communication noise. Postprocessing approaches for more accurate estimation are described in Phase III.

4.2.3   STEP III: POSTPROCESSING

The state vectors seen in Section 4.2.2 do not converge perfectly to the degree distribution due to noisy communications and lack of convergence in finite time. In this section, we introduce two postprocessing methods to make the estimation more accurate.

In the first postprocessing method, a threshold is used to decide whether the convergence result is purely noise or not. For node i, the estimate of the degree distribution at time t^* can be expressed as

$$\hat{p}_i^{(k)}(t^*) = \begin{cases} \mathbf{x}_i^{(k)}(t^*), & \text{if } \mathbf{x}_i^{(k)}(t^*) \geq \frac{1}{2N} \\ 0, & \text{otherwise.} \end{cases} \tag{4.5}$$

Note that the threshold is chosen based on the network size, N in Equation (4.5).

The second method is to postprocess the convergence result based on the fact that the degree values are always integers. We assume that the number of nodes in the network, N, is given. Assuming that the algorithm stops at time t^*, the postprocessing step can be expressed as,

$$\hat{p}_i^{(k)}(t^*) = \frac{\lfloor N \hat{p}_i^{(k)}(t^*) \rceil}{N}, \tag{4.6}$$

where $\lfloor x \rceil$ rounds x to the nearest integer. More accurate degree distribution estimation can be obtained by applying the postprocessing steps mentioned above as will be shown in the simulations. A pseudo code for the three steps of the proposed distributed network degree distribution estimation algorithm is given in Algorithm 4.1.

Note that the size of the state vector, K, can be reduced. A simple way to reduce the size of the state vector can be to let each bin cover the same degree ranges, for example 1 to 2, 3 to 4, 5 to 6, and so on. The final estimate can be obtained by dividing the result of each element in the state vector by the width of the bin to normalize the measurement. In practice, some networks such as random networks, have power-law degree distributions. Therefore, one way to reduce the size of the state vector is to let bins cover an increasing range of degrees, for example 1, 2 to 3, 4 to 7, 8 to 15, and so on [146].

4.3   ESTIMATION OF DEGREE MATRIX

Assume that consensus is reached and the degree distribution is obtained at all nodes by using the proposed algorithm in Section 4.2. Then, if the total number of nodes in the network N is known, the degree matrix can be calculated. The estimate of number of nodes with degree k can be calculated as

$$\hat{n}_k = \left\lfloor N \hat{p}_i^{(k)} \right\rceil, \tag{4.7}$$

where $\lfloor x \rceil$ rounds x to the nearest integer. Therefore, we know there are $\hat{n}_k$ nodes with degree k and $k = 1, 2, \cdots K$. Then the diagonal degree matrix can be obtained and there are totally $\hat{n}_k$ diagonal elements with value k, $k = 1, 2, \cdots K$. Note that since nodes are not labeled, the obtained

---

Algorithm 4.1 Distributed Degree Distribution Estimation

Generate Initial Values
for each node i:

$$\mathbf{x}_i^{(k)}(0) = \begin{cases} 1, & \text{if } d_i = k \\ 0, & \text{otherwise,} \end{cases}$$

$$k = 1, 2, \cdots, K.$$

Iterative Average Consensus
for each node i:

$$\mathbf{x}_i^{(k)}(t+1) = [1 - \alpha(t)d_i]\,\mathbf{x}_i^{(k)}(t) + \alpha(t) \sum_{j \in \mathbb{N}_i} \left[\mathbf{x}_j^{(k)}(t) + v_{ij}^{(k)}(t) \right],$$

until predefined stopping time t^* is reached.

Postprocessing
for each node i:
degree distribution estimation at i at time t^*:

$$\hat{p}_i^{(k)}(t^*) = \begin{cases} \mathbf{x}_i^{(k)}(t^*), & \text{if } \mathbf{x}_i^{(k)}(t^*) \geq \frac{1}{2N} \\ 0, & \text{otherwise.} \end{cases}$$

---

degree matrix is actually a permutation of the labeled degree matrix. Note that $\sum_k n_k = N$, where n_k is the actual number of nodes with degree k.

4.4   PERFORMANCE ANALYSIS

In this section, the performance of the degree distribution estimation algorithm is analyzed: (i) The convergence speed of the algorithm is given in Section 4.4, and (ii) how the communication noise affects the performance is provided in Section 4.4.

Convergence Speed Analysis

Since average consensus is used, the performance analysis for average consensus in [64] is used to quantify the convergence speed for the degree distribution estimation. Let $\mathbf{x}^{(k)}(t) = \left[x_1^{(k)}(t) \cdots x_N^{(k)}(t) \right]^T$ contain all the kth element in the state vectors from all nodes at iteration time t; the convergence speed of the algorithm is characterized by the error between the expected estimated result and the true degree distribution. We have

$$||\mathbf{E}[\mathbf{x}^{(k)}(t)] - \frac{n_k}{N}\mathbf{1}||_2 \leq \left(\prod_{0<\tau\leq t}(1-\alpha(\tau)\lambda_2(\mathbf{L}))\right)||\mathbf{x}^{(k)}(0) - \frac{n_k}{N}\mathbf{1}||_2. \qquad (4.8)$$

We have the following conclusions based on Equation (4.8): (1) The convergence rate of the proposed degree distribution estimation algorithm depends on the network connectivity $\lambda_2(\mathbf{L})$, the step size $\alpha(t)$, and the initial state values $\mathbf{x}^{(k)}(0)$; and (2) the expected value of the states of nodes, $\mathbf{E}[\mathbf{x}^{(k)}(t)]$ converge exponentially to the degree distribution when $(1-\alpha(\tau)\lambda_2(\mathbf{L})) < 1$.

Steady State Error Analysis
In this section, we assume that the convergence is reached and how the communication noise affects the steady state convergence result is analyzed. The distribution of the steady state convergence result is derived, and we also show how the communication noise and design parameters affect the distribution of the convergence result (estimated degree distribution).

Assume consensus is reached as $t \to \infty$; the kth element in the state vector of node i can be expressed as,

$$\mathbf{x}_i^{(k)}(t) = \frac{n_k}{N} + v', \qquad (4.9)$$

where v' is the accumulated communication noise. Since average consensus is used and assume noise is Gaussian distributed, we have,

$$v' \sim \mathcal{N}\left(0, \left(\frac{\sum_{i=1}^{N}d_i}{N^2}\right)\sigma_n^2\sum_{t=0}^{\infty}\alpha^2(t)\right). \qquad (4.10)$$

Therefore, $\mathbf{x}_i^{(k)}(t)$ is also Gaussian distributed: we have,

$$\mathbf{x}_i^{(k)}(t) \sim \mathcal{N}\left(\frac{n_k}{N}, \left(\frac{\sum_{i=1}^{N}d_i}{N^2}\right)\left(\frac{\sigma_n^2}{1}\right)\sum_{t=0}^{\infty}\alpha^2(t)\right). \qquad (4.11)$$

We have the following based on Equation (4.11): (i) The convergence result is an unbiased estimator of the degree distribution; and, (ii) smaller noise variance results in smaller variance for $\mathbf{x}_i^{(k)}(t)$. Therefore we have more accurate degree distribution estimation. Note that a more detailed performance analysis is given in [147].

4.5   SIMULATIONS

In this section, simulation results for the proposed algorithms are presented. The network topology is the same as in Figure 3.2. The true degree distribution is given in Figure 4.2, where the x-axis values are possible values of degree, $k = 1, 2, \cdots, K$ and the y-axis is the probability $\Pr[X = k]$.

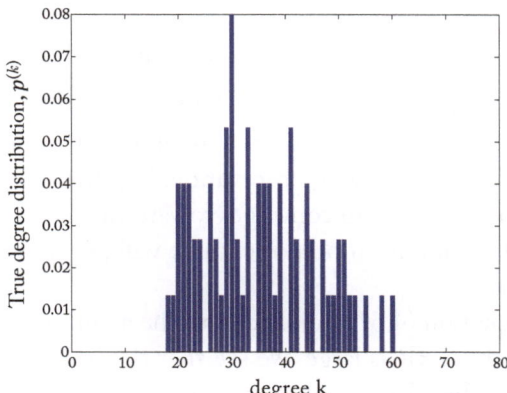

Figure 4.2: True degree distribution.

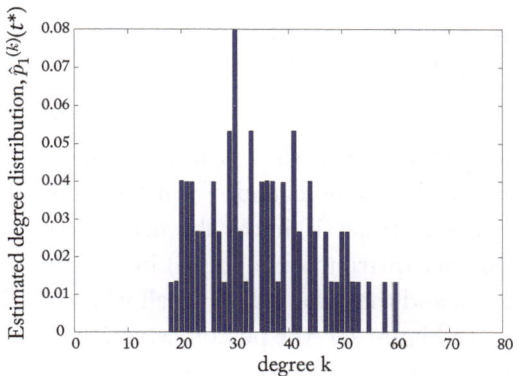

Figure 4.3: Estimate of degree distribution at time $t^* = 100$ at node 1 in the absence of noise, $\sigma_n^2 = 0$ and $\alpha(t) = 0.1/t$.

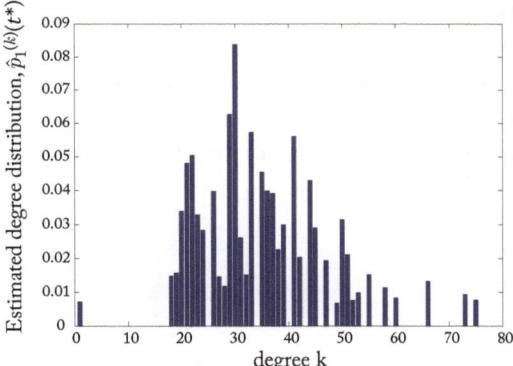

Figure 4.4: Estimate of degree distribution at time $t^* = 100$ at node 1 in the presence of noise, $\sigma_n^2 = 0.1$ and $\alpha(t) = 0.1/t$.

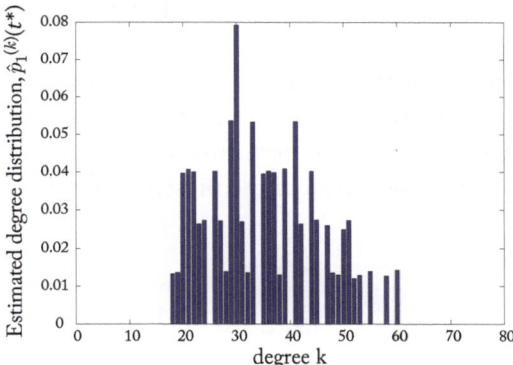

Figure 4.5: Estimate of degree distribution at time $t^* = 100$ at node 1 in the presence of noise, $\sigma_n^2 = 0.01$ and $\alpha(t) = 0.1/t$.

The proposed algorithm with threshold postprocessing is performed and results are shown in Figures 4.3–4.5. We assume that the algorithm stops at $t^* = 100$, and set $\alpha(t) = 0.1/t$, and $K = 80$. In Figure 4.3, we assume perfect communication with no communication noise. From the results, we can see that an accurate estimate of the degree distribution can be obtained.

Results with noisy channels are given in Figure 4.4 and Figure 4.5. The noise variance is $\sigma_n^2 = 0.1$ in Figure 4.4 and $\sigma_n^2 = 0.01$ in Figure 4.5. By comparing the two figures, we can conclude that the proposed algorithm converges toward a reasonable estimate of the network degree distribution in the presence of communication noise and more accurate degree distribution estimation can be obtained at higher SNR.

In Figure 4.6, the error norm of the consensus result at node 1, $||\mathbf{x}_1(t) - \mathbf{p}||$ is shown, where $\mathbf{p}$ is the true degree distribution and $\mathbf{x}_1(t) = \left[\mathbf{x}_1^{(1)}(t), \cdots, \mathbf{x}_1^{(K)}(t) \right]$ is the degree distribution estimate at node 1 at time t. Noise variances σ_n^2 are chosen to have different values, as shown in the figure legend. From Figure 4.6 we have the following observations: (i) The error decreases vs. time; (ii) in the absence of noise, the consensus result converges toward the desired true degree distribution; and (iii) in the presence of communication noise, there will always be an error, and the error will be small when the SNR is large.

In Figure 4.7, the postprocessing based on Equation (4.6) is applied with the assumption that the number of nodes $N = 75$ is given. When the SNR is large, we see that the line with circle terminates after six iterations, which indicates that the error becomes exactly 0 after six iterations if we postprocess the convergence result as in Equation (4.6). However, when the SNR is small, the postprocessing does not improve the performance much, as shown in Figure 4.7 (compared the two curves with $\sigma_n^2 = 0.1$).

To conclude, we have the following observations based on the simulations: (1) Accurate degree distribution can be obtained by running the proposed distributed degree distribution estimation algorithm in the absence of communication noise; (2) in the presence of noise, reasonable estimate of the degree distribution can also be obtained; and (3) the proposed postprocessing methods improve the performance of the algorithm, and accurate degree distribution estimation can be obtained even in the presence of noise (when SNR is large).

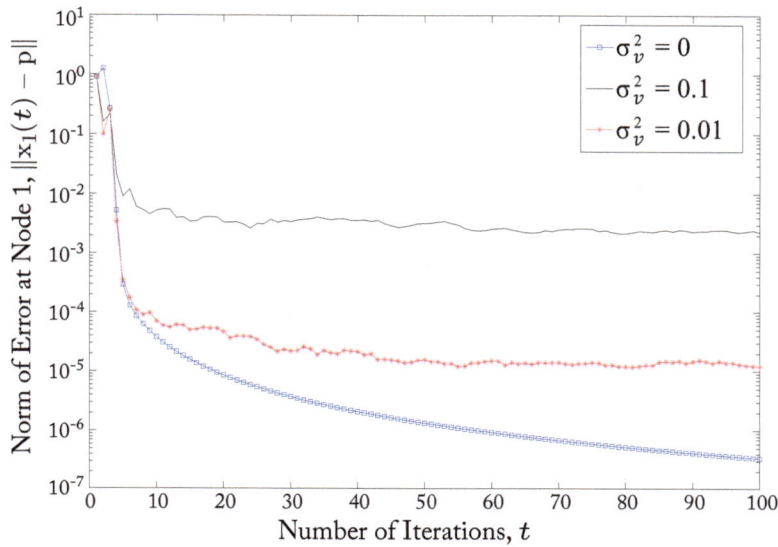

Figure 4.6: Error vs. t, $\alpha(t) = 0.1/t$ for noise variance $\sigma_n^2 = 0, 0.01, 0.01$.

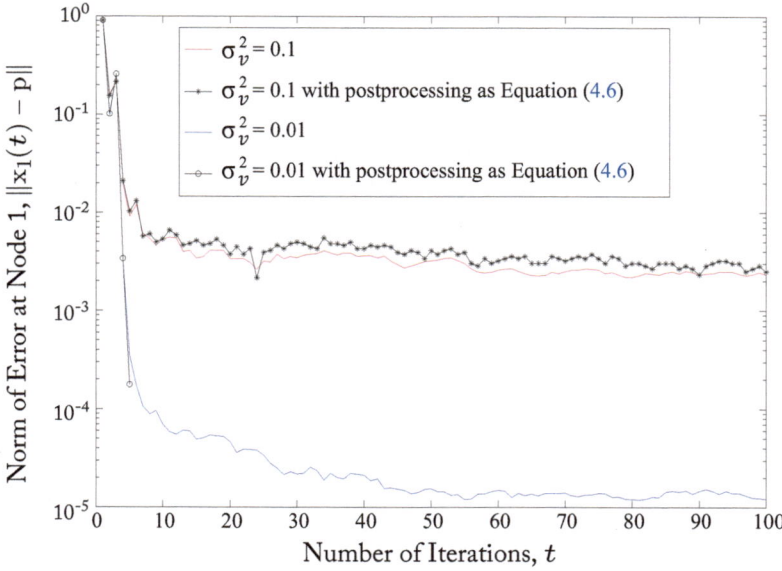

Figure 4.7: Simulation results for postprocessing as in Equation (4.6): error vs. t. The line with circle terminates after 6 iterations, which indicates that the error becomes exactly 0 after six iterations with proposed postprocessing.

CHAPTER 5

Network Center and Coverage Region Estimation

In this chapter, a fully distributed algorithm for estimating the center and the coverage region of a distributed WSN is introduced. The network coverage region is defined to be the smallest covering sphere. Therefore, the center and radius characterize the coverage region. Note that part of the content in this chapter is presented in [136, 148, 149].

Knowing the network coverage region and center is useful in many applications. Network center information used for localizing a service center is given in [90]. Applications of using network coverage region and area information for deciding optimal connections and energy-efficient scheduling are given in [86, 91].

5.1 SYSTEM MODEL

A connected fully distributed WSN with no fusion center is considered and the network is modeled as a graph as described previously in Section 2.1. It is assumed that each node in the network only knows its own location. Each sensor always keeps a single state and the sensors update their states by exchanging their states with their neighbors. We assume that there is no communication noise between nodes.

A 2-D network with center at the origin and radius equal to 1 is given in Figure 5.1.

5.2 ESTIMATION OF NETWORK CENTER AND RADIUS

The main idea of the algorithm is to use soft-max approximation to formulate the center estimation problem as a convex optimization problem with a summation form. Then distributed optimization methods such as stochastic gradient descent and diffusion adaptation can be used to estimate the center. After all nodes obtain the estimate of the center, max consensus is used for distributed radius calculation and the network coverage region can be obtained at nodes. In the following, details of the center and radius estimation algorithms are presented.

5.2.1 DISTRIBUTED CENTER ESTIMATION

For a spherical network coverage region, the network center, ω^*, is defined to be the point that minimizes the maximum distance between ω^* and all nodes [90]. In the 2-D case, network

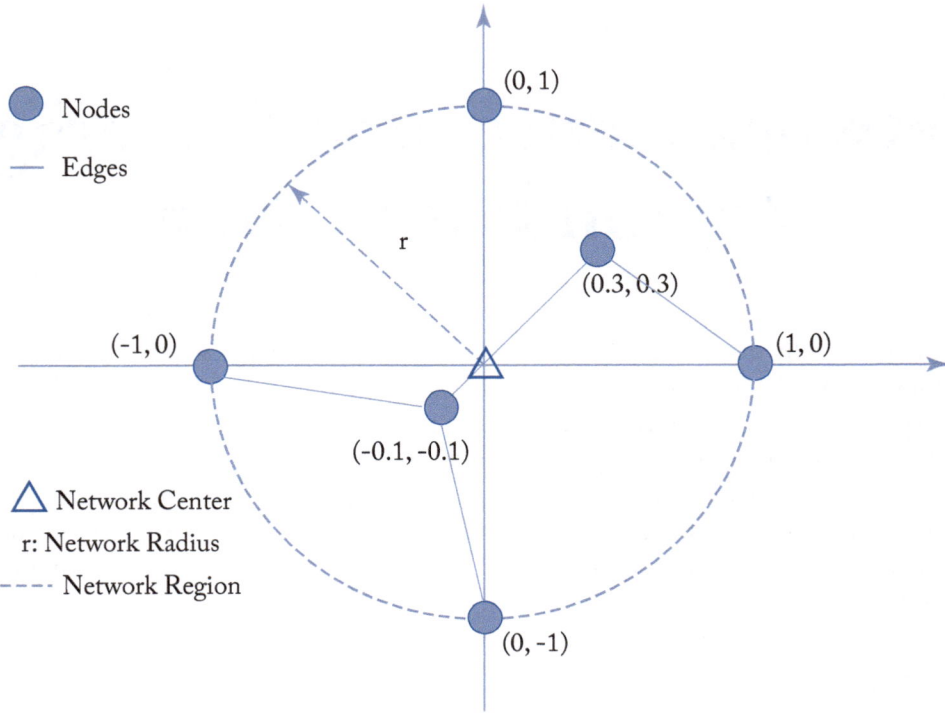

Figure 5.1: A distributed network (2-D) with $N = 6$ nodes. A circle is used to represent the network coverage region.

center estimation can be posed as the following optimization problem:

$$\omega^* = \text{argmin}_{x,y} \; \max_i \left\{ (a_i - x)^2 + (b_i - y)^2 \right\},\tag{5.1}$$

where (a_i, b_i) is the location of node i and the location of the center $\omega^* = [x^* \; y^*]$ minimizes (5.1). In the 3-D case, the optimization problem can be written as

$$\omega^* = \text{argmin}_{x,y,z} \; \max_i \left\{ (a_i - x)^2 + (b_i - y)^2 + (c_i - z)^2 \right\},\tag{5.2}$$

where (a_i, b_i, c_i) is the location of node i and the location of the center $\omega^* = [x^* y^* z^*]$ minimizes (5.2). In the following, we will focus on the 2-D case. The algorithm for the 3-D case is similar.

The objective function in Equation (5.1) is neither differentiable nor convex. However, soft-max approximation can be used to reformulate the optimization problem in Equation (5.1). The soft-max function can be used to approximate the maximum. The soft maximum of a vector

$\boldsymbol{\theta} = [\theta_1 \; \theta_2 \; \cdots \; \theta_N]$ is denoted as

$$\text{smax}(\boldsymbol{\theta}) = \frac{1}{\beta} \log \sum_{i=1}^{N} e^{\beta \theta_i}, \tag{5.3}$$

where $\beta > 0$ is a design parameter and the soft-max approximates the maximum for large β.

Equation (5.1) can be formulated as the following optimization problem using the soft-max approximation:

$$\omega_\beta^* = \text{argmin}_{x,y} \; \frac{1}{\beta} \log \left[\sum_{i=1}^{N} e^{\beta \{(a_i - x)^2 + (b_i - y)^2\}} \right]. \tag{5.4}$$

Note that ω^* and ω_β^* in Equations (5.1) and (5.4) will be the same when $\beta \to \infty$. Also note that the value of ω_β^* depends on the value of β. Equation (5.4) can be further simplified since $\log(\cdot)$ is a monotonically increasing function; we have

$$\omega_\beta^* = \text{argmin}_{x,y} \; \sum_{i=1}^{N} e^{\beta \{(a_i - x)^2 + (b_i - y)^2\}}. \tag{5.5}$$

Therefore, by using soft-max approximation, the center estimation problem is formulated as a convex optimization problem, and the objective function is in the form of the sum of locally differentiable functions. The proof of convexity of Equation (5.5) is obtained by first calculating the Hessian of the objective function in Equation (5.5). Then we can show that the Hessian matrix is positive definite when $\beta > 0$. The details of the proof are given in [136].

The property of convexity and the summation form of Equation (5.5) enable us to use distributed optimization algorithms to estimate the network center. In what follows, two different distributed center estimation algorithms for solving Equation (5.5) are described: the stochastic gradient descent method and the diffusion adaptation method.

Center Estimation Using Stochastic Gradient Descent

Since the objective function in Equation (5.5) is a sum of differentiable functions, stochastic gradient method can be used to solve the convex optimization problem in a fully distributed manner [140, 150].

Algorithm 5.2 presents the update steps at nodes for estimating the network center. Note that $(x_i(t), y_i(t))$ is the state of node i at time t, η is the gradient descent step size, $J_i(x_i(t), y_i(t))$ is the local objective function at iteration time time t, and

$$J_i(x_i(t), y_i(t)) = e^{\beta \{(a_i - x_i(t))^2 + (b_i - y_i(t))^2\}}. \tag{5.6}$$

In Algorithm 5.2, firstly, one of the sensor nodes is selected as a leader, with starting value $(x(0), y(0))$. The leader updates its estimate using stochastic gradient descent method and, randomly choosing one of its neighbors, passes the estimate to the chosen node. The node that receives the data becomes an active node and the original leader turns inactive. Then the active node: (i) updates its state $(x_i(t), y_i(t))$ by using stochastic gradient descent and randomly chooses one neighboring node to pass the estimate to; and (ii) after passing the data, the original source node becomes inactive and the node that gets the data becomes active. Note that at each iteration time, there is only one active node in the network doing the update, and all the inactive nodes stay idle. Finally, when t is large, all nodes reach consensus on the estimated center.

Note that Algorithm 5.2 performs like a sequential stochastic gradient descent algorithm, and it is fully distributed in the sense that the update of $x_i(t + 1)$ and $y_i(t + 1)$ at node i only depends on its own location information (a_i, b_i) and received data $(x_i(t), y_i(t))$. Also note that max consensus can be used for distributed leader selection (choose the starting node) at the beginning of Algorithm 5.2 [81].

Algorithm 5.2 Stochastic gradient descent for center calculation

---

select a leader node (active node), with starting values $(x(0), y(0))$.

for active node i:

  repeat

    $x_i(t + 1) = x_i(t) - \eta \frac{\partial}{\partial x_i(t)} J_i(x_i(t), y_i(t))$.

    $y_i(t + 1) = y_i(t) - \eta \frac{\partial}{\partial y_i(t)} J_i(x_i(t), y_i(t))$.

    select a neighbor $j \in \mathcal{N}_i$ to pass data:

    $x_j(t + 1) = x_i(t + 1), \ y_j(t + 1) = y_i(t + 1)$.

    node $i \rightarrow$ inactive, node $j \rightarrow$ active

  until stopping criterion is satisfied.

Center Estimation Using Diffusion Adaptation

Since the global cost function in Equation (5.5) is in the form of summation of individual real-valued local functions, and the local functions are differentiable and convex, diffusion adaptation strategies like those in [70, 151] can be used to achieve consensus on the network center estimate.

Let $\boldsymbol{\omega}_i(t) = [x_i(t) \ y_i(t)]^{\mathrm{T}}$, and $\boldsymbol{\psi}_i(t) \in \mathbb{R}^2$ be intermediate value vectors. The diffusion adaptation iterative updating algorithm can be expressed as:

$$\boldsymbol{\psi}_i(t + 1) = \boldsymbol{\omega}_i(t) - \mu \sum_{j \in \mathbb{N}_i} c_{j,i} \nabla_{\boldsymbol{\omega}} J_j(\boldsymbol{\omega}_i(t)), \tag{5.7}$$

$$\boldsymbol{\omega}_i(t + 1) = \sum_{j \in \mathbb{N}_i} a_{j,i} \boldsymbol{\psi}_i(t + 1), \tag{5.8}$$

where $\mu > 0$ is the descent step size, and $c_{j,i}$ and $a_{j,i}$ are nonnegative coefficients that satisfy

$$\sum_{i=1}^{N} c_{j,i} = 1, c_{j,i} = 0 \text{ if } j \notin \mathbb{N}_i, j = 1, 2, \cdots, N, \tag{5.9}$$

$$\sum_{i=1}^{N} a_{j,i} = 1, a_{j,i} = 0 \text{ if } j \notin \mathbb{N}_i, j = 1, 2, \cdots, N. \tag{5.10}$$

At each iteration time, the algorithm involves two steps. In the first step in Equation (5.7), node i updates the intermediate vector $\psi_i(t + 1)$ based on its own estimate $\omega_i(t)$ and the gradient vector information from its neighbors. In the second step in Equation (5.8), nodes exchange information with their neighbors and calculate the network center, $\omega_i^{(t+1)}$, based on the received intermediate results.

The diffusion adaptation algorithm in Equations (5.7) and (5.8) and the stochastic gradient descent algorithm (Algorithm 5.2) both have their advantages and disadvantages. The diffusion adaptation algorithm usually converges faster than the stochastic gradient descent approach since all nodes are active and exchanging information with their neighbors at each iteration time. However, the diffusion algorithm is more complicated and requires that all nodes are synchronized and update at the same time [8, 151].

5.2.2   DISTRIBUTED RADIUS ESTIMATION

After all nodes reach consensus on the estimated center (x^*, y^*), max consensus can be used for radius calculation. Each node first sets its initial state to be its distance to the estimated center. The initial state for node i, denoted as $r_i(0)$, can be expressed as

$$r_i(0) = \sqrt{(a_i - x^*)^2 + (b_i - y^*)^2}. \tag{5.11}$$

Then max consensus using the max operator as described in Section 2.2.2 is used for estimating the radius in the network. The updating rule is straightforward: The nodes update their states with the largest received measurement they receive in each iteration. Let $r_i(t)$ be the state of node i at time t; the updating rule can be expressed as

$$r_i(t + 1) = \max \left\{ r_i(t), \max_{j \in \mathbb{N}_i} r_j(t) \right\}, \tag{5.12}$$

where $r_j(t)$ with $j \in \mathbb{N}_i$ are the states of neighbors of node i. It is proved in [80] and [82] that by running the iterative algorithm, the states of the nodes converge to the maximum of the initial states in finite time. Therefore, the radius is obtained when consensus is reached,

$$\hat{r} = \max_i \{r_1(0) \ r_2(0) \ \cdots r_i(0) \ \cdots r_N(0)\} \tag{5.13}$$

$$= \max_i \sqrt{(a_i - x^*)^2 + (b_i - y^*)^2}. \tag{5.14}$$

Therefore, by knowing the center and radius, all nodes in the network obtain an estimate of the network coverage region.

5.3   PERFORMANCE ANALYSIS

In this section, the steady state performance analysis of the network center estimation algorithm described in Section 5.2.1 is given. We assume that consensus is reached and all sensor nodes converge to the estimate of the center. The accuracy of the center estimation depends on the design parameter β and the locations of nodes (a_i, b_i).

Steady state performance of the center estimation algorithm depends on the design parameter β since soft-max approximation function is used. The approximation from Equation (5.1) to Equation (5.5) is accurate when β is large. Therefore, more accurate network center estimates can be obtained at nodes when β is chosen to be large.

We also have an interesting finding that the performance of center estimation also depends on the locations of nodes (a_i, b_i). It is shown that when the locations of nodes are symmetric around the center, an accurate center estimation can be obtained even when β is small. The proof and analysis are given in the following. For simplicity, we consider the simple $1 - D$ case in this part with nodes locations at a_i, and results can be easily extended to higher dimensional cases. The center of all nodes depends on the maximum and minimum nodes locations and is at location $\omega^* = x^* = (a_{\max} + a_{\min})/2$. By using the soft-max approximation formulation as in Section 5.2, the steady state center location is calculated as:

$$\text{argmin}_x \sum_{i=1}^{N} e^{\beta(a_i - x)^2}. \tag{5.15}$$

The optimal solution to Equation (5.15), x_{β}^* can be found by taking derivative of Equation (5.15) and set to 0. Therefore, x_{β}^* can be calculated as:

$$\sum_{i=1}^{N} (x_{\beta}^* - a_i) e^{\beta(x_{\beta}^* - a_i)^2} = 0. \tag{5.16}$$

We have the following observations based on Equation (5.16): (i) the estimated center x_{β}^* is related to the initial nodes locations a_i; and (ii) when the nodes locations are symmetric around the center, the solution of Equation (5.16), x_{β}^*, will be the same as the true network center, $x^* = (a_{\max} + a_{\min})/2$. This is because when locations are symmetric, we can assume without loss of generality that $x^* - a_i = a_{N-i+1} - x^*$. Assume N is even, we have:

$$\sum_{i=1}^{N}(x^* - a_i)e^{\beta(x^*-a_i)^2} \tag{5.17}$$

$$= \sum_{i=1}^{N/2}(x^* - a_i)e^{\beta(x^*-a_i)^2} + \sum_{i=\frac{N}{2}+1}^{N}(x^* - a_i)e^{\beta(x^*-a_i)^2} \tag{5.18}$$

$$= \sum_{i=1}^{N/2}(x^* - a_i)e^{\beta(x^*-a_i)^2} - \sum_{i=1}^{N/2}(x^* - a_i)e^{\beta(x^*-a_i)^2} = 0. \tag{5.19}$$

As shown in Equations (5.17)–(5.19), x^* satisfies Equation (5.16). Therefore, symmetric node locations leads to accurate steady state center estimation, regardless of the value of β. Note that examples of asymptotic symmetric node locations are uniformly distributed nodes.

Note that the proposed center estimation algorithm is based on the stochastic gradient descent and diffusion adaptation method. The convergence speed of stochastic gradient descent and diffusion adaptation has been studied extensively in the literature [8, 151, 152] and can be used to analyze the convergence speed of the proposed center estimation algorithm. Also note that the proposed radius estimation algorithm in Section 5.2.2 always converges in finite iterations since max consensus with the max operator is used. The convergence speed of max consensus depends on the minimum number of edges needed to connect any two nodes in the network graph. Details of the convergence speed for max consensus can be found in [80].

5.4   SIMULATIONS

In this section, simulation results for the algorithm described in this chapter are presented. A 2-D connected graph with $N = 6$ nodes is considered, and the locations of nodes are shown in Figure 5.2. The center of the network is at $(0, 0)$ and the radius is $r = 1$.

In Figures 5.3 and 5.4, stochastic gradient descent-based Algorithm 5.2 is performed for estimating the x and y coordinates of the center. We set node 1 (location at $(1, 0)$) to be the starting node with starting values $(0.3, 0.8)$. The stochastic gradient descent step size is $\eta = 10^{-4}$ and the soft-max parameter is $\beta = 1$. From the results, we can see that the estimate converges to the center of the network when t is large.

In Figures 5.5 and 5.6, diffusion adaptation in Equations (5.7) and (5.8) is used for distributed center estimation. The x and y coordinate estimates at different iteration times t are plotted. The initial states at nodes $\omega_i(0) = [x_i(0) \; y_i(0)]^\mathrm{T}$ are set to be uniformly distributed, $\mathcal{U}(-0.5, 0.5)$. The descent step size for diffusion adaptation based method is $\mu = 10^{-4}$ and the soft-max parameter $\beta = 1$. The coefficients $c_{j,i}$ and $a_{j,i}$ are set based on the degree of nodes:

$$c_{j,i} = a_{j,i} = \begin{cases} \frac{1}{d_i+1}, & \text{if } j \in \mathbb{N}_i \\ 0, & \text{otherwise.} \end{cases} \tag{5.20}$$

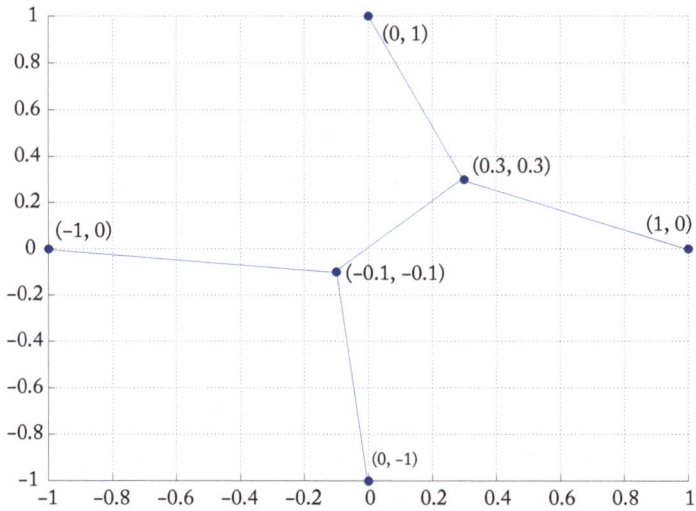

Figure 5.2: Graph representation of the sensor network, $N = 6$.

We can see from the figure that the estimates converge to the center.

The center estimation error for the two distributed estimation algorithms are plotted in Figures 5.7 and 5.8. In Figure 5.7, Algorithm 5.2 is used and the error between the center O and the estimated center at node 1, denoted as $\sqrt{(x_1(t) - x_O)^2 + (y_1(t) - y_O)^2}$, is plotted, where $(x_O, y_O) = (0, 0)$ is the center. In Figure 5.8, diffusion adaptation method is performed and the average estimation error vs. iteration time t is plotted. From Figures 5.8 and 5.7, we can conclude that the error decreases as t increases and the estimates converge toward the center. We can also see from the figures that the method of diffusion adaptation converges faster than the stochastic gradient descent.

Max consensus is performed for radius estimation. We assume that the stochastic gradient descent method is used and the iterative algorithm stops at $t^* = 5000$. Distances from nodes to the estimated centers are calculated at nodes and set as initial values at nodes for max consensus. Figure 5.9 shows the max consensus radius estimation results. We can see from the figure that consensus is reached in three iterations and the estimated radius $\hat{r} = 1.063$.

Finally based on the estimated center and radius, estimate of the network coverage region is obtained at nodes. The estimated network coverage region of node 1 at time $t^* = 5000$ is shown in Figure 5.10.

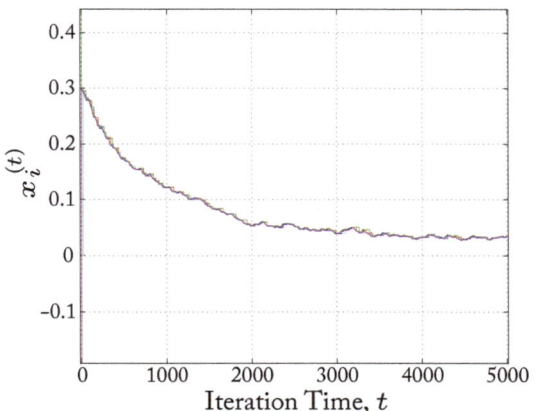

Figure 5.3: Estimate of the x coordinate value of the center, $x_i(t)$ vs. iteration t using Algorithm 5.2, $\eta = 10^{-4}$, and starting point $x^{(0)} = 0.3$.

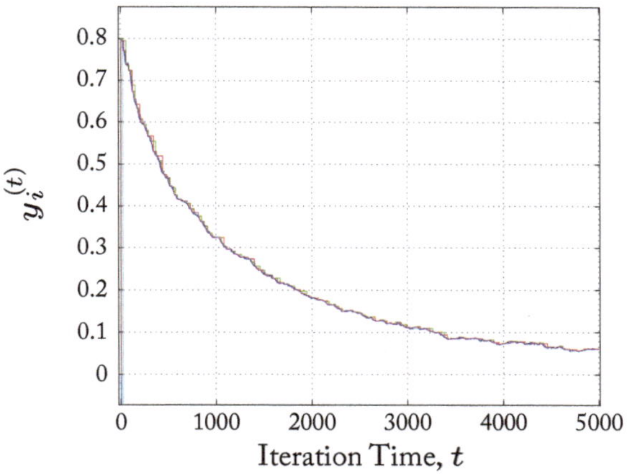

Figure 5.4: Estimate of the y coordinate value of the center, $y_i(t)$ vs. iteration t using Algorithm 5.2, $\eta = 10^{-4}$, and starting point $y^{(0)} = 0.8$.

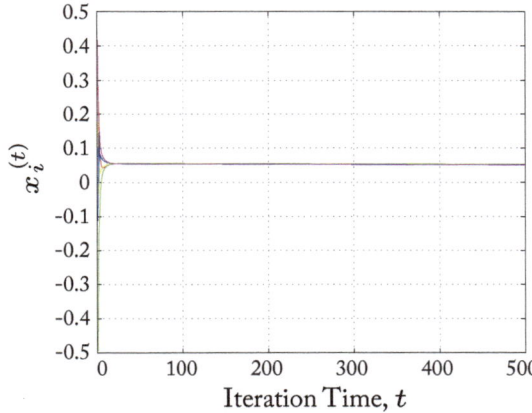

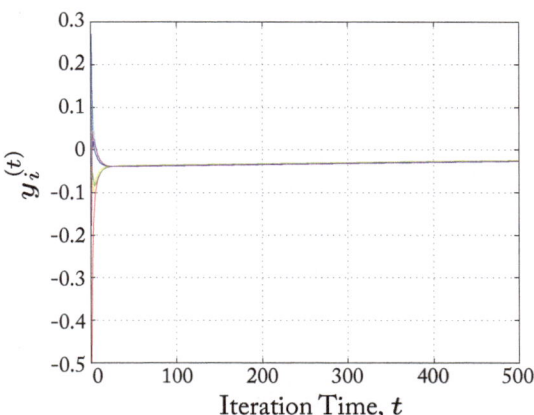

Figure 5.5: Estimate of the x coordinate value of the center, $x_i(t)$ vs. iteration t using diffusion adaptation, $\eta = 10^{-4}$, and starting point to be uniformly distributed $\mathcal{U}(-0.5, 0.5)$.

Figure 5.6: Estimate of the y coordinate value of the center, $y_i(t)$ vs. iteration t using diffusion adaptation, $\eta = 10^{-4}$, and starting point to be uniformly distributed $\mathcal{U}(-0.5, 0.5)$.

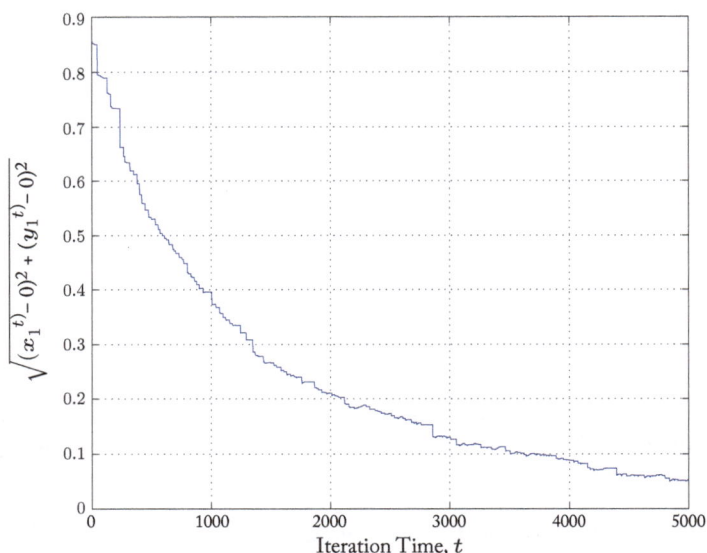

Figure 5.7: Error vs. t at node 1 with the Algorithm 5.2, where $O(x_O, y_O)$ is the true center and $x_O = 0$, $y_O = 0$.

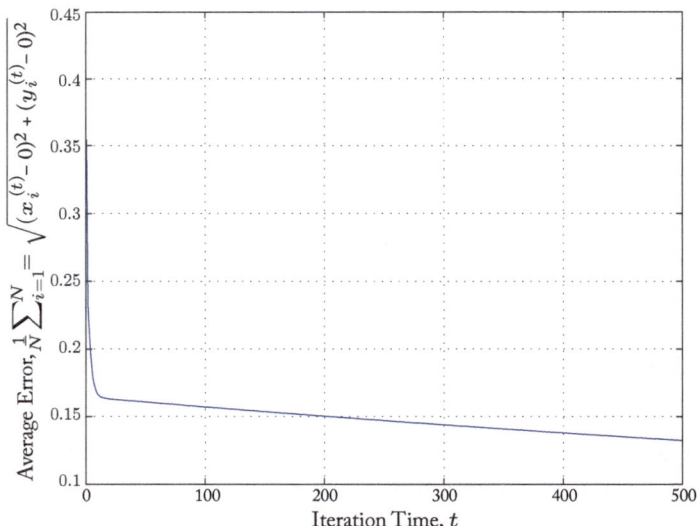

Figure 5.8: Average error vs. t using diffusion adaptation, where $O(x_O, y_O)$ is the true center and $x_O = 0$, $y_O = 0$.

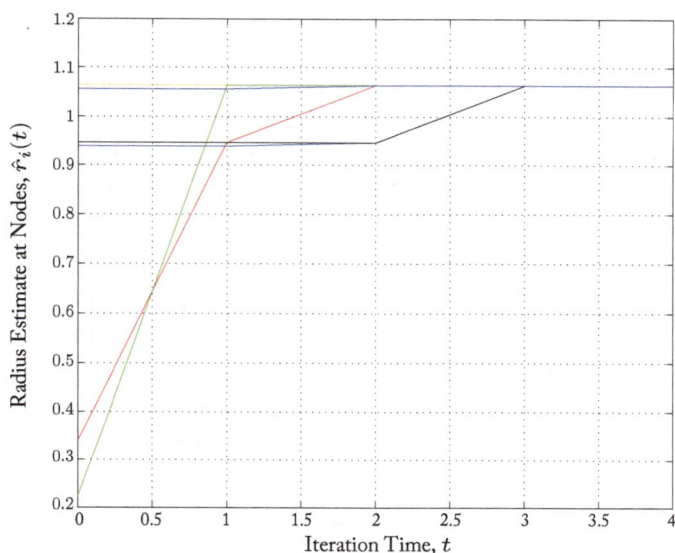

Figure 5.9: Radius estimate vs. t using max consensus in Section 5.2.2. The initial value at node i is set to be the distance between the estimated center and its own location.

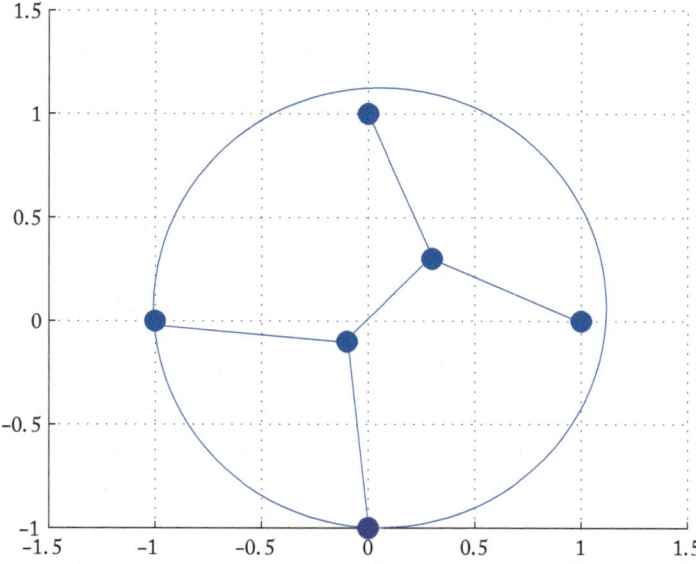

Figure 5.10: Estimated network coverage region at node 1 at $t = 5000$.

5.5   DISCUSSION: GLOBAL DATA STRUCTURE ESTIMATION

The algorithm for estimating the center and coverage region of sensor nodes can also be used for estimating the center and region/size of the global data at nodes in a distributed network. The algorithm is useful for letting nodes in distributed network learn the global data information and "visualize" the global data structure. For estimating the center and region of global data, we assumed that each node in the network has a $K-$dimensional initial data. It is desired that by running the iterative algorithm, all the nodes reach consensus on an estimate of the center and the smallest $K-$ dimensional sphere that covers all data points from nodes.

The iterative algorithm introduced in Section 5.2.1 can be used for estimating the center of data points with slight modification. Let $\alpha_i \in \mathbb{R}^K$ be the $K-$dimensional initial data at node i. By applying the soft-max approximation in Equation (5.3), the center of the data, ω^*, can be estimated by solving the following convex optimization problem (approximation steps are the same as in Section 5.2.1):

$$\text{argmin}_{\omega} \sum_{i=1}^{N} e^{\beta \|\alpha_i - \omega\|_2^2}. \tag{5.21}$$

The convexity and summation form in Equation (5.21) enable us to use the distributed iterative update step as in Equations (5.7) and (5.8) to estimate the data center. In the algorithm, the estimate of the data center at node i at time t is denoted as $\omega_i(t)$, and $\psi_i(t)$ is a $K-$dimensional auxiliary vector at node i. At each iteration, each node i first updates $\psi_i(t+1)$ based on the estimated center from previous iteration $\omega_i(t)$ and the gradient vector information from its neighbors as in Equation (5.7). Then nodes exchange information with their neighbors and combine the intermediate results to calculate the estimated data center as in Equation (5.8). By running the above iterative update steps, nodes will reach consensus on an estimated global data center.

After the estimated data center is obtained at all nodes, the max consensus update in Section 5.2.2 is used to estimate the radius of the global data at nodes. Note that the initial states for max consensus are set to be the norm $\|\alpha_i - \omega^*\|_2$. Finally the global data size and structure information can be inferred and visualized from the center and radius information.

A simulation example of the distributed data center and the global data region information estimation is given in Figures 5.12 and 5.13. The connectivity structure of the distributed network is as shown in Figure 5.11 with $N = 75$. Each sensor has a initial data vector, and the data at node i is $[a_i \ b_i \ c_i]$. In the simulation, the values of data at nodes are fixed and are generated as below: eight data vectors are at "vertex" $(0,0,0)$, $(0,0,1)$, $(0,1,0)$, $(1,0,0)$, $(1,1,0)$, $(1,0,1)$, $(0,1,1)$, $(1,1,1)$, and the rest are generated from uniform distribution, $a_i, b_i, c_i \sim \mathcal{U}(0,1)$, therefore the global data center is at $(0.5, 0.5, 0.5)$ and the radius of data $r = 0.866$. The estimate of the data center at nodes vs. time is given in Figure 5.12, and the estimate of the global data region at node 1 is shown in Figure 5.14. We observe in the simulations that by running the iterative algorithm, nodes in the network reach consensus on a reasonable estimate of the data center and region. Therefore, they learn the global data information.

Note that the other algorithms introduced in this book, such as the distributed degree distribution estimation algorithm in Section 4, can also be used for estimating the distribution of data and learning the structure of data at nodes in a distributed way.

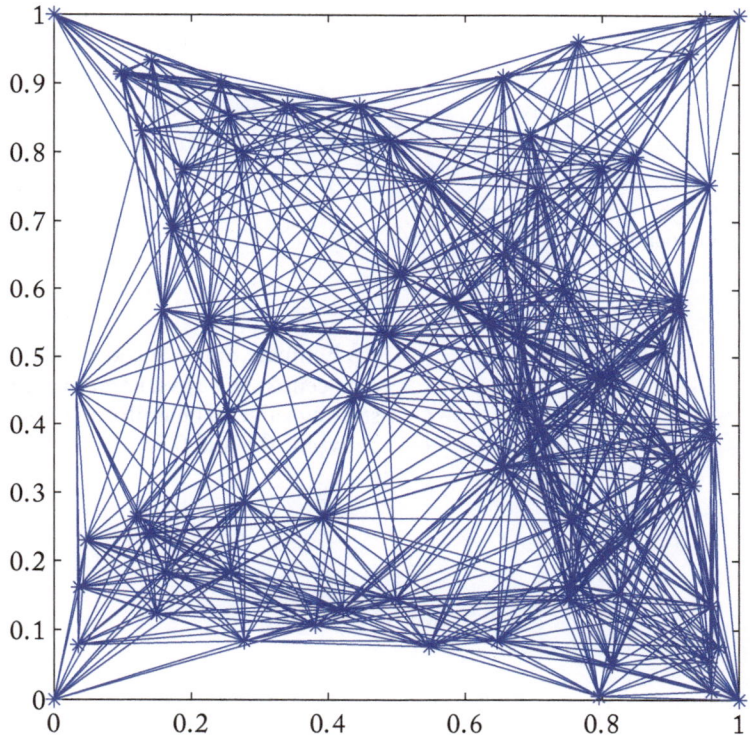

Figure 5.11: Graph representation of a network with $N = 75$.

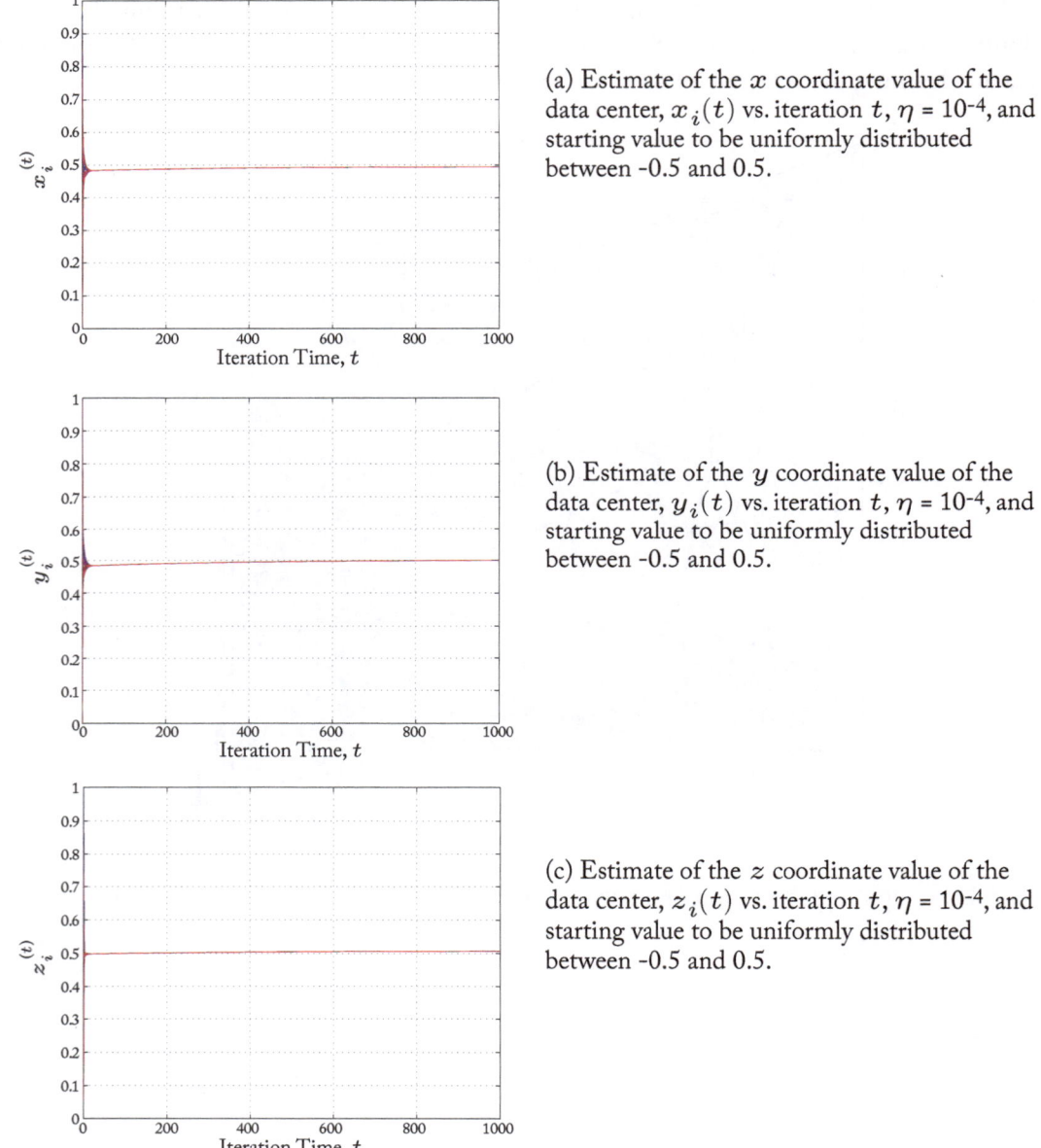

(a) Estimate of the x coordinate value of the data center, $x_i(t)$ vs. iteration t, $\eta = 10^{-4}$, and starting value to be uniformly distributed between -0.5 and 0.5.

(b) Estimate of the y coordinate value of the data center, $y_i(t)$ vs. iteration t, $\eta = 10^{-4}$, and starting value to be uniformly distributed between -0.5 and 0.5.

(c) Estimate of the z coordinate value of the data center, $z_i(t)$ vs. iteration t, $\eta = 10^{-4}$, and starting value to be uniformly distributed between -0.5 and 0.5.

Figure 5.12: Estimate of the 3-D data center at all nodes as described in Section 5.5.

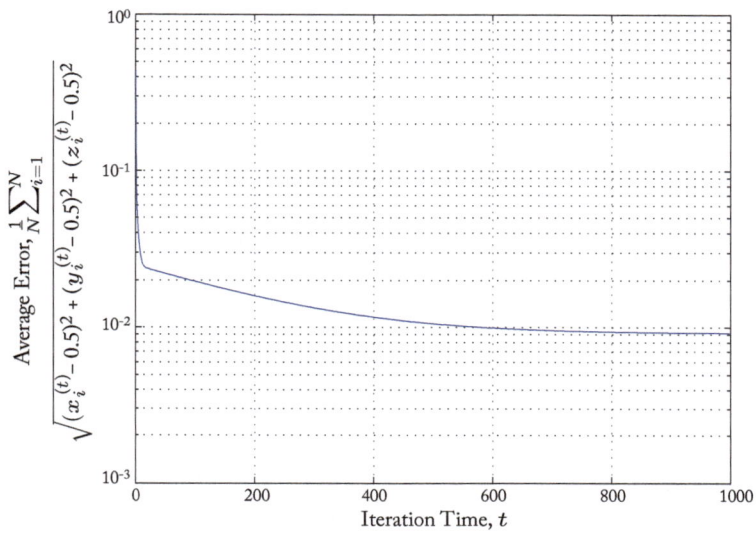

Figure 5.13: Average error vs. t for data center estimation described in Section 5.5, where $(0.5, 0.5, 0.5)$ is the true center of data.

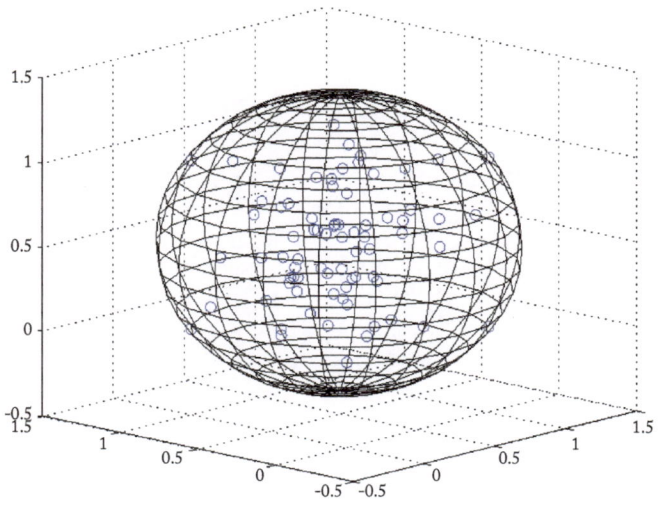

Figure 5.14: Estimated global data region at node 1.

CHAPTER 6

Conclusions

This book covers distributed wireless sensor networks structure estimation. In Chapter 1, the basic concepts of wireless sensor networks including consensus estimation algorithms and related topics and applications are introduced. The mathematical background of methods used in consensus-based WSN framework is reviewed in Chapter 2. A brief review on graph theory is given, and existing distributed consensus methods and algorithms for estimating the structure of network are described.

In Chapter 3, we describe a distributed consensus algorithm for estimating the system size, in other words, estimating the total number of nodes in a wireless sensor network with noisy communication channels. The idea of average consensus and L_2 norm approximation is described and different sources of error are explicitly discussed. The performance of the final estimate at each node is also derived and discussed in detail. We introduce several WSN design parameters and discuss how they affect the performance of distributed consensus estimation. An optimal design parameter that minimizes the L_2 norm estimation error is derived. We also demonstrate the trade off between estimation accuracy and the storage at the nodes. We show that a more accurate system size estimation can be obtained if we increase the storage at the sensor nodes.

In Chapter 4, we present a practical approach for reliable estimation of the degree distribution in a distributed WSN in the presence of noise. The main idea in this framework is that the degrees of a WSN are discrete values and an average consensus with postprocessing can be applied to estimate the degree distribution. The properties of the degree distribution are used to postprocess the consensus results in order to obtain more accurate estimates. We also show that the number of bins that are used to represent the degree distribution can be reduced in many practical situations, which results in storage saving at the sensor nodes.

In Chapter 5, we presented a fully distributed network center and network coverage region estimation algorithm. The coverage region is defined to be the smallest sphere that covers the entire set of the sensor nodes, and the center and radius of the network coverage region are estimated. The key idea is that by using a soft-max approximation, the conventional center estimation optimization problem can be formulated as a convex optimization problem. Moreover, the objective function can be written in the form of summation of local real-valued convex functions. The summation form and the convexity property enable us to use distributed optimization methods such as stochastic gradient descent and diffusion adaptation to solve the problem of center estimation in a distributed manner. Finally, max consensus is used to compute the radius.

Therefore, the network area and coverage region can be obtained at the nodes. We show that reasonable estimates of the network center and coverage region can be obtained approximately in a finite number of iterations. We also show that the proposed algorithm is useful for learning the global data information in a distributed network.

This book serves as an introduction to distributed consensus algorithms and distributed structure estimation in wireless sensor networks. Further reading is needed to gain a full appreciation of the field of distributed WSN structure estimation. For further reading on network connectivity, coverage, and related applications we suggest [153–155]. For graph sampling and known sub-graphs for use in network structure estimation, the reader can refer to [153, 156–158], and for the WSNs node or data clustering problems, we cite [12, 94, 159, 160]. Finally, we note several broad studies that engage machine-learning algorithms in sensors and in WSNs [24, 161–168].

APPENDIX A

Notation

$\mathbb{G}$	wireless sensor network graph
$\mathbb{N}$	set of wireless sensor nodes
$\mathbb{E}$	set of network edges
N	number of sensor nodes in network
i, j	sensor node index
t	iteration time index
d_i	degree of sensor node i
$\mathbf{A}$	graph adjacency matrix
$\mathbf{D}$	graph degree matrix
$\mathbf{L}$	graph Laplacian matrix
$\mathbf{I}$	identity matrix
$\mathbf{W}$	weight matrix
$\mathbf{x}$	state vector
$x_i(t)$	state of node i at iteration time t
$\rho(\cdot)$	spectral radius of matrix
n	communication noise
σ_n^2	noise variance
$\alpha, \alpha(t)$	iterative algorithm step size
$E[\cdot]$	expected value
$\mathrm{Var}[\cdot]$	variance
$[\ldots]^T$	vector or matrix transpose
$\hat{\mathbf{x}}$	estimate of state $\mathbf{x}$
$\lVert \cdot \rVert_2$	vector L_2 norm
$\bar{\cdot}$	average; mean
$\lfloor \cdot \rceil$	round
$p(\cdot)$	probability density function (pdf)
$\mathcal{N}$	Gaussian distribution
$\mathcal{U}$	Uniform distribution
$\mathbf{J}_i(\cdot)$	local objective function at node i

Bibliography

[1] I. Akyildiz, W. Su, Y. Sankarasubramaniam, and E. Cayirci, A survey on sensor networks, *IEEE Communications Magazine*, 40, 2002. DOI: 10.1109/mcom.2002.1024422. 1

[2] J. A. Stankovic, Wireless sensor networks, *Computer*, 41(10), pp. 92–95, October 2008. DOI: 10.1109/mc.2008.441.

[3] C.-Y. Chong and S. P. Kumar, Sensor networks: Evolution, opportunities, and challenges, *Proc. of the IEEE*, 91(8), pp. 1247–1256, August 2003. DOI: 10.1109/jproc.2003.814918.

[4] D. Culler, D. Estrin, and M. Srivastava, Overview of sensor networks, *Computer*, 37(8), pp. 41–49, August 2004. DOI: 10.1109/mc.2004.93. 1

[5] Wikipedia, Wireless sensor network—Wikipedia, the free encyclopedia, 2014. https://en.wikipedia.org/wiki/Wireless_sensor_network 1

[6] J. Yick, B. Mukherjee, and G. Dipak, Wireless sensor network survey, *Computer Networks*, 52, pp. 2292–2330, August 2008. DOI: 10.1016/j.comnet.2008.04.002. 1, 2, 5

[7] M. Goldenbaum, S. Stanczak, and M. Kaliszan, On function computation via wireless sensor multiple-access channels, *Wireless Communications and Networking Conference*, pp. 1–6, April 2009. DOI: 10.1109/wcnc.2009.4917843. 1, 4, 11, 15

[8] F. Cattivelli and A. Sayed, Diffusion LMS strategies for distributed estimation, *IEEE Transactions on Signal Processing*, 58, pp. 1035–1048, 2010. DOI: 10.1109/tsp.2009.2033729. 1, 4, 43, 45

[9] O. Younis and S. Fahmy, Distributed clustering in ad-hoc sensor networks: A hybrid, energy-efficient approach, *IEEE INFOCOM*, 1, p. 640, March 2004. DOI: 10.1109/infcom.2004.1354534.

[10] A. G. Dimakis, S. Kar, J. M. F. Moura, M. G. Rabbat, and A. Scaglione, Gossip algorithms for distributed signal processing, *Proc. of the IEEE*, 98(11), pp. 1847–1864, November 2010. DOI: 10.1109/jproc.2010.2052531.

[11] J. F. Chamberland and V. V. Veeravalli, Decentralized detection in sensor networks, *IEEE Transactions on Signal Processing*, 51(2), pp. 407–416, February 2003. DOI: 10.1109/tsp.2002.806982. 1

60 BIBLIOGRAPHY

[12] R. K. Shakya, Y. N. Singh, and N. K. Verma, A novel spatial correlation model for wireless sensor network applications, *9th International Conference on Wireless and Optical Communications Networks (WOCN)*, pp. 1–6, September 2012. DOI: 10.1109/wocn.2012.6335549. 1, 56

[13] R. W. Santucci, M. K. Banavar, A. Spanias, and C. Tepedelenlioglu, Design of limiting amplifier models for nonlinear amplify-and-forward distributed estimation, *18th International Conference on Digital Signal Processing (DSP)*, pp. 1–6, July 2013. DOI: 10.1109/icdsp.2013.6622749. 1, 4

[14] R. W. Santucci, M. K. Banavar, A. Spanias, and C. Tepedelenlioglu, Nonlinear amplify and forward distributed estimation over non-identical channels, *IEEE Transactions on Vehicular Technologies*, 64(11), pp. 5390–5395, November 2015. DOI: 10.1109/tvt.2014.2381094. 4

[15] M. Krämer, S. Bader, and B. Oelmann, Implementing wireless sensor network applications using hierarchical finite state machines, *10th IEEE International Conference on Networking, Sensing and Control (ICNSC)*, pp. 124–129, April 2013. DOI: 10.1109/icnsc.2013.6548723.

[16] K. Ghosal, T. Anand, V. Chaturvedi, and B. Amrutur, A power-scalable RF CMOS receiver for 2.4 GHz wireless sensor network applications, *19th IEEE International Conference on Electronics, Circuits, and Systems (ICECS)*, pp. 161–164, December 2012. DOI: 10.1109/icecs.2012.6463775.

[17] U. Pesovic, Z. Jovanovic, S. Randjic, and D. Markovic, Benchmarking performance and energy efficiency of microprocessors for wireless sensor network applications, *Proc. of the 35th International Convention MIPRO*, pp. 743–747, May 2012. 1

[18] R. Zeng and C. Tepedelenlioglu, Multiple device-to-device users overlaying cellular networks, *IEEE Wireless Communications and Networking Conference (WCNC)*, pp. 1–6, March 2017. DOI: 10.1109/wcnc.2017.7925800. 1

[19] I. Akyildiz, W. Su, Y. Sankarasubramaniam, and E. Cayirci, Wireless sensor networks: A survey, *Computer Networks*, 38, pp. 393–422, 2002. DOI: 10.1002/9780470515181. 1

[20] S. Zhang, C. Tepedelenlioglu, M. Banavar, and A. Spanias, Max consensus in sensor networks: Non-linear bounded transmission and additive noise, *IEEE Sensors Journal*, 16, pp. 9089–9098, December 2016. DOI: 10.1109/jsen.2016.2612642. 5, 15, 16, 17

[21] R. Saber and R. Murray, Consensus protocols for networks of dynamic agents, *Proc. of the American Control Conference*, pp. 951–956, June 2003. DOI: 10.1109/acc.2003.1239709. 1, 11

[22] T. Arampatzis, J. Lygeros, and S. Manesis, A survey of applications of wireless sensors and wireless sensor networks, *Proc. of the IEEE International Symposium on, Mediterrean Conference on Control and Automation Intelligent Control*, pp. 719–724, June 2005. DOI: 10.1109/.2005.1467103. 1

[23] U. Prathap, P. D. Shenoy, K. R. Venugopal, and L. M. Patnaik, Wireless sensor networks applications and routing protocols: Survey and research challenges, *International Symposium on Cloud and Services Computing*, pp. 49–56, December 2012. DOI: 10.1109/iscos.2012.21. 2, 5

[24] U. Shanthamallu, A. Spanias, C. Tepedelenlioglu, and M. Stanley, A brief survey of machine learning methods and their sensor and IoT applications, *Proc. 8th International Conference on Information, Intelligence, Systems and Applications (IEEE IISA)*, August 2017. 1, 3, 5, 56

[25] C. Meesookho, S. Narayanan, and C. S. Raghavendra, Collaborative classification applications in sensor networks, *Sensor Array and Multichannel Signal Processing Workshop Proceedings*, pp. 370–374, August 2002. DOI: 10.1109/sam.2002.1191063. 1, 5

[26] D. Li, K. D. Wong, Y. H. Hu, and A. M. Sayeed, Detection, classification, and tracking of targets, *IEEE Signal Processing Magazine*, 19(2), pp. 17–29, March 2002. DOI: 10.1109/79.985674. 1

[27] M. Hammoudeh, F. Al-Fayez, H. Lloyd, R. Newman, B. Adebisi, A. Bounceur, and A. Abuarqoub, A wireless sensor network border monitoring system: Deployment issues and routing protocols, *IEEE Sensors Journal*, 17(8), pp. 2572–2582, April 2017. DOI: 10.1109/jsen.2017.2672501. 1, 2, 6

[28] E. Felemban, Advanced border intrusion detection and surveillance using wireless sensor network technology, *International Journal of Communications, Network and System Sciences*, 6, pp. 251–259, May 2013. DOI: 10.4236/ijcns.2013.65028. 2, 5, 6

[29] M. K. Banavar, J. J. Zhang, B. Chakraborty, H. Kwone, Y. Li, H. Jiang, A. Spanias, C. Tepedelenlioglu, C. Chakrabartie, and A. Papandreou-Suppappola, An overview of recent advances on distributed and agile sensing algorithms and implementation, *Digital Signal Processing*, 39, pp. 1–14, April 2015. DOI: 10.1016/j.dsp.2015.01.001. 2, 5

[30] S. Peter and P. Langendörfer, Tool-supported methodology for component-based design of wireless sensor network applications, *IEEE 36th Annual Computer Software and Applications Conference Workshops*, pp. 526–531, July 2012. DOI: 10.1109/compsacw.2012.98.

[31] H. Braun, C. Tepedelenlioglu, A. Spanias, and M. Banavar, *Signal Processing for Solar Array Monitoring, Fault Detection, and Optimization, Synthe-*

sis Lectures on Power Electronics, Morgan & Claypool Publishers, 2012. DOI: 10.2200/S00425ED1V01Y201206PEL004. 2

[32] M. Castillo-Effer, D. H. Quintela, W. Moreno, R. Jordan, and W. Westhoff, Wireless sensor networks for flash-flood alerting, *Proc. of the 5th IEEE International Caracas Conference on Devices, Circuits and Systems*, 1, pp. 142–146, November 2004. DOI: 10.1109/iccdcs.2004.1393370. 2

[33] G. Werner-Allen, K. Lorincz, M. Ruiz, O. Marcillo, J. Johnson, J. Lees, and M. Welsh, Deploying a wireless sensor network on an active volcano, *IEEE Internet Computing*, 10(2), pp. 18–25, March 2006. DOI: 10.1109/mic.2006.26.

[34] K. Lorincz, D. J. Malan, T. R. F. Fulford-Jones, A. Nawoj, A. Clavel, V. Shnayder, G. Mainland, M. Welsh, and S. Moulton, Sensor networks for emergency response: Challenges and opportunities, *IEEE Pervasive Computing*, 3(4), pp. 16–23, October 2004. DOI: 10.1109/mprv.2004.18. 2, 5

[35] M. K. Banavar, C. Tepedelenlioglu, and A. Spanias, Distributed snr estimation with power constrained signaling over gaussian multiple-access channels, *IEEE Transactions on Signal Processing*, 60, pp. 3289–3294, February 2012. DOI: 10.1109/tsp.2012.2188524. 2, 5

[36] N. Kovvali, M. Banavar, and A. Spanias, *An Introduction to Kalman Filtering with MATLAB Examples*, Synthesis Lectures on Signal Processing, Morgan & Claypool Publishers, 2013. DOI: 10.2200/s00534ed1v01y201309spr012. 2, 5

[37] Y. Zhou and S. Maskell, RB^2-PF : A novel filter-based monocular visual odometry algorithm, *20th International Conference on Information Fusion*, IEEE, pp. 1–8, July 2017. DOI: 10.23919/icif.2017.8009745.

[38] Y. Wang, Y. Sheng, J. Wang, and W. Zhang, Optimal collision-free robot trajectory generation based on time series prediction of human motion, *IEEE Robotics and Automation Letters*, 3(1), pp. 226–233, January 2018. DOI: 10.1109/lra.2017.2737486. 2, 5

[39] M. Usman, V. Muthukkumarasamy, X. W. Wu, and S. Khanum, Securing mobile agent based wireless sensor network applications on middleware, *International Symposium on Communications and Information Technologies (ISCIT)*, pp. 707–712, October 2012. DOI: 10.1109/iscit.2012.6380993. 2

[40] A. Simonetto, T. Keviczky, and D. V. Dimarogonas, Distributed solution for a maximum variance unfolding problem with sensor and robotic network applications, *50th Annual Allerton Conference on Communication, Control, and Computing (Allerton)*, pp. 63–70, October 2012. DOI: 10.1109/allerton.2012.6483200. 2, 5

[41] X. Zhang, C. Tepedelenlioglu, M. Banavar, and A. Spanias, *Node Localization in Wireless Sensor Networks, Synthesis Lectures on Communications*, Morgan & Claypool Publishers, 2016. DOI: 10.2200/s00742ed1v01y201611com012.

[42] X. Zhang, M. Banavar, C. Tepedelenlioglu, and A. Spanias, Maximum likelihood localization in the presence of channel uncertainties, November 2016, US Patent. `https://www.google.com/patents/US9507011?hl=zh-CN#backward-citations` 2, 5

[43] J. Lee, M. Stanley, A. Spanias, and C. Tepedelenlioglu, Integrating machine learning in embedded sensor systems for internet-of-things applications, *IEEE International Symposium on Signal Processing and Information Technology (ISSPIT)*, pp. 290–294, Limassol, December 2016. DOI: 10.1109/isspit.2016.7886051. 2

[44] C. Tepedelenlioglu, M. K. Banavar, and A. Spanias, On the asymptotic efficiency of distributed estimation systems with constant modulus signals over multiple-access channels, *IEEE Transactions on Information Theory*, 57(10), pp. 7125–7130, October 2011. DOI: 10.1109/tit.2011.2165806. 2, 5

[45] X. Zhang, C. Tepedelenlioglu, M. K. Banavar, and A. Spanias, Distributed location detection in wireless sensor networks, *Asilomar Conference on Signals, Systems and Computers*, pp. 428–432, November 2013. DOI: 10.1109/acssc.2013.6810312.

[46] X. Zhang, M. K. Banavar, M. Willerton, A. Manikas, C. Tepedelenlioglu, A. Spanias, T. Thornton, E. Yeatman, and A. G. Constantinides, Performance comparison of localization techniques for sequential WSN discovery, *Sensor Signal Processing for Defence (SSPD)*, pp. 1–5, September 2012. DOI: 10.1049/ic.2012.0120. 6, 7

[47] J. Foutz, A. Spanias, and M. Banavar, *Narrowband Direction of Arrival Estimation for Antenna Arrays, Synthesis Lectures on Antennas*, Morgan & Claypool Publishers, 2008. DOI: 10.2200/s00118ed1v01y200805ant008. 2, 5

[48] Y. H. Nam, Z. Halm, Y. J. Chee, and K. S. Park, Development of remote diagnosis system integrating digital telemetry for medicine, *Proc. of the 20th Annual International Conference of the IEEE*, pp. 1170–1173, October 1998. DOI: 10.1109/iembs.1998.747079. 2, 5

[49] D. Shao, Y. Yang, C. Liu, F. Tsow, H. Yu, and N. Tao, Noncontact monitoring breathing pattern, exhalation flow rate and pulse transit time, *IEEE Transactions on Biomedical Engineering*, 61(11), pp. 2760–2767, November 2014. DOI: 10.1109/tbme.2014.2327024.

[50] T. Gao, D. Greenspan, M. Welsh, R. R. Juang, and A. Alm, Vital signs monitoring and patient tracking over a wireless network, *IEEE Engineering in Medicine and Biology 27th Annual Conference*, pp. 102–105, January 2005. DOI: 10.1109/iembs.2005.1616352.

[51] M. Mathew, N. Weng, and L. J. Vespa, Quality-of-information modeling and adapting for delay-sensitive sensor network applications, *IEEE 31st International Performance Computing and Communications Conference (IPCCC)*, pp. 471–477, December 2012. DOI: 10.1109/pccc.2012.6407659.

[52] G. Fortino, R. Giannantonio, R. Gravina, P. Kuryloski, and R. Jafari, Enabling effective programming and flexible management of efficient body sensor network applications, *IEEE Transactions on Human-machine Systems*, 43(1), pp. 115–133, January 2013. DOI: 10.1109/tsmcc.2012.2215852. 2

[53] D. Shao, Y. Yang, F. Tsow, C. Liu, and N. Tao, Non-contact simultaneous photoplethysmogram and ballistocardiogram video recording towards real-time blood pressure and abnormal heart rhythm monitoring, *12th IEEE International Conference on Automatic Face Gesture Recognition (FG)*, pp. 273–277, May 2017. DOI: 10.1109/fg.2017.42. 2

[54] D. Shao, F. Tsow, C. Liu, Y. Yang, and N. Tao, Simultaneous monitoring of ballistocardiogram and photoplethysmogram using a camera, *IEEE Transactions on Biomedical Engineering*, 64(5), pp. 1003–1010, May 2017. DOI: 10.1109/tbme.2016.2585109.

[55] C. R. Baker, K. Armijo, S. Belka, M. Benhabib, V. Bhargava, N. Burkhart, A. D. Minassians, G. Dervisoglu, L. Gutnik, M. B. Haick, C. Ho, M. Koplow, J. Mangold, S. Robinson, M. Rosa, M. Schwartz, C. Sims, H. Stoffregen, A. Waterbury, E. S. Leland, T. Pering, and P. K. Wright, Wireless sensor networks for home health care, *Advanced Information Networking and Applications Workshops, (AINAW), 21st International Conference on*, 2, pp. 832–837, May 2007. DOI: 10.1109/AINAW.2007.376. 2, 5

[56] S. Rao, D. Ramirez, H. Braun, J. Lee, C. Tepedelenlioglu, E. Kyriakides, D. Srinivasan, J. Frye, S. Koizumi, Y. Morimoto, and A. Spanias, An 18 kw solar array research facility for fault detection experiments, *18th Mediterranean Electrotechnical Conference (MELECON)*, pp. 1–5, Limassol, April 2016. DOI: 10.1109/melcon.2016.7495369. 3, 5

[57] S. Rao, S. Katoch, A. Spanias, C. Tepedelenlioglu, R. Ayyanar, H. Braun, J. Lee, U. Shanthamallu, M. Banavar, and D. Srinivasan, A cyber-physical system approach for photovoltaic array monitoring and control, *Proc. 8th International Conference on Information, Intelligence, Systems and Applications (IEEE IISA)*, Larnaca, August 2017.

[58] A. Spanias, Solar energy management as an internet of things (IoT) applications, *Proc. 8th International Conference on Information, Intelligence, Systems and Applications (IEEE IISA)*, Larnaca, August 2017. 3

[59] H. Braun, S. Buddha, V. Krishnan, C. Tepedelenlioglu, A. Spanias, M. Banavar, and D. Srinivansan, Topology reconfiguration for optimization of photovoltaic array output, *Elsevier Sustainable Energy, Grids and Networks*, 6, pp. 58–69, June 2016. DOI: 10.1016/j.segan.2016.01.003. 3, 5, 6

[60] N. A. Lynch, *Distributed Algorithms*, Morgan Kaufmann, 1997. 3

[61] R. Olfati-Saber, J. A. Fax, and R. M. Murray, Consensus and cooperation in networked multi-agent systems, *IEEE Signal Processing Magazine*, 95(1), 2007. DOI: 10.1109/jproc.2006.887293. 3, 4, 11

[62] J. Baillieul and P. J. Antsaklis, Control and communication challenges in networked real-time systems, *Proc. of the IEEE*, 95(1), pp. 9–28, January 2007. DOI: 10.1109/jproc.2006.887290. 3

[63] W. Ren, R. W. Beard, and E. M. Atkins, A survey of consensus problems in multi-agent coordination, *Proc. of the American Control Conference*, 3, pp. 1859–1864, June 2005. DOI: 10.1109/acc.2005.1470239.

[64] S. Kar and J. Moura, Distributed consensus algorithms in sensor networks with imperfect communication: Link failures and channel noise, *IEEE Transactions on Signal Processing*, 57(1), pp. 355–369, January 2009. DOI: 10.1109/tsp.2008.2007111. 3, 4, 12, 13, 21, 24, 31, 33

[65] S. Kar and J. M. F. Moura, Sensor networks with random links: Topology design for distributed consensus, *IEEE Transactions on Signal Processing*, 56(7), pp. 3315–3326, July 2008. DOI: 10.1109/tsp.2008.920143. 3

[66] R. Olfati-Saber and R. Murray, Consensus problems in networks of agents with switching topology and time-delays, *IEEE Transactions on Automatic Control*, 49(9), pp. 1520–1533, September 2004. DOI: 10.1109/tac.2004.834113. 3, 4, 9, 11, 14

[67] L. Xiao, S. Boyd, and S. Kim, Distributed average consensus with least-mean-square deviation, *Journal of Parallel and Distributed Computing*, 67, pp. 33–46, 2007. DOI: 10.1016/j.jpdc.2006.08.010. 3, 12, 16

[68] R. Santucci, M. Banavar, C. Tepedelenlioglu, and A. Spanias, Energy-efficient distributed estimation by utilizing a nonlinear amplifier, *IEEE Transactions on Circuits and Systems I: Regular Papers*, pp. 302–311, January 2014. DOI: 10.1109/tcsi.2013.2268354. 1, 3, 4

[69] L. Xiao and S. Boyd, Fast linear iterations for distributed averaging, *Proc. 42nd IEEE Conference on Decision and Control*, 5, pp. 4997–5002, December 2003. DOI: 10.1109/cdc.2003.1272421. 4, 5, 11, 12, 16, 21

[70] J. Lee, C. Tepedelenlioglu, M. K. Banavar, and A. Spanias, Nonlinear diffusion adaptation with bounded transmission over distributed networks, *IEEE ICC*, pp. 6707–6711, London, June 2015. DOI: 10.1109/icc.2015.7249394. 3, 42

[71] A. Papachristodoulou, A. Jadbabaie, and U. Munz, Effects of delay in multi-agent consensus and oscillator synchronization, *IEEE Transactions on Automatic Control*, 55(6), pp. 1471–1477, 2010. DOI: 10.1109/tac.2010.2044274. 4, 11

[72] R. Olfati-Saber, Flocking for multi-agent dynamic systems: Algorithms and theory, *IEEE Transaction on Automatic Control*, 51(3), pp. 401–420, March 2006. DOI: 10.1109/tac.2005.864190. 4

[73] R. Olfati-Saber and R. M. Murray, Distributed cooperative control of multiple vehicle formations using structural potential functions, *IFAC Proceedings Volumes, 15th IFAC World Congress*, 35(1), pp. 495–500, 2002. http://www.sciencedirect.com/science/article/pii/S1474667015386651 DOI: 10.3182/20020721-6-es-1901.00244. 4

[74] R. Olfati-Saber and R. M. Murray, Graph rigidity and distributed formation stabilization of multi-vehicle systems, *Proc. of the 41st IEEE Conference on Decision and Control*, 3, pp. 2965–2971, December 2002. DOI: 10.1109/cdc.2002.1184307. 4

[75] L. Xiao, S. Boyd, and S. Lall, A scheme for robust distributed sensor fusion based on average consensus, *Fourth International Symposium on Information Processing in Sensor Networks*, pp. 63–70, April 2005. DOI: 10.1109/ipsn.2005.1440896. 4, 5, 11

[76] K. C. Chang, C. Y. Chong, and Y. Bar-Shalom, Distributed estimation in distributed sensor networks, *Large-scale Stochastic Systems Detection, Estimation, Stability and Control*, S. G. Tzafestas and K. Watanabe, Eds., ch. 2, pp. 23–71, Marcel Dekker, 1992. 4

[77] D. Zhao, Z. An, and Y. Xu, Time synchronization in wireless sensor networks using max and average consensus protocol, *International Journal of Distributed Sensor Networks*, February 2013. DOI: 10.1155/2013/192128. 4, 11

[78] S. Dasarathan, C. Tepedelenlioglu, M. Banavar, and A. Spanias, Non-linear distributed average consensus using bounded transmissions, *IEEE Transactions on Signal Processing*, 61, pp. 6000–6009, December 2013. DOI: 10.1109/tsp.2013.2282912. 4, 12, 13, 14, 16, 21, 24, 31

[79] S. Dasarathan, C. Tepedelenlioglu, M. Banavar, and A. Spanias, Robust consensus in the presence of impulsive channel noise, *IEEE Transactions on Signal Processing*, 63, pp. 2118–2129, March 2015. DOI: 10.1109/tsp.2015.2408564. 4, 13

[80] F. Iutzeler, P. Ciblat, and J. Jakubowicz, Analysis of max-consensus algorithms in wireless channels, *IEEE Transactions on Signal Processing*, 60, pp. 6103–6107, November 2012. DOI: 10.1109/tsp.2012.2211593. 4, 14, 43, 45

[81] A. Tahbaz-Salehi and A. Jadbabaie, A one-parameter family of distributed consensus algorithms with boundary: From shortest paths to mean hitting times, *45th*

IEEE Conference on Decision and Control, pp. 4664–4669, December 2006. DOI: 10.1109/cdc.2006.377308. 15, 42

[82] B. Nejad, S. Attia, and J. Raisch, Max-consensus in a max-plus algebraic setting: The case of fixed communication topologies, *International Symposium on Information, Communication and Automation Technologies*, pp. 1–7, October 2009. DOI: 10.1109/icat.2009.5348437. 14, 15, 43

[83] G. Shi and K. H. Johansson, Convergence of distributed averaging and maximizing algorithms Part II: State-dependent graphs, *American Control Conference*, pp. 6859–6864, June 2013. DOI: 10.1109/acc.2013.6580916. 15

[84] S. Giannini, A. Petitti, D. D. Paola, and A. Rizzo, Asynchronous max-consensus protocol with time delays: Convergence results and applications, *IEEE Transactions on Circuits and Systems I: Regular Papers*, 63, pp. 256–264, January 2016. DOI: 10.1109/tcsi.2015.2512721. 4, 14

[85] S. Zhang, C. Tepedelenlioglu, M. Banavar, and A. Spanias, Max-consensus using the soft maximum, *Asilomar Conference on Signals, Systems and Computers*, pp. 433–437, November 2013. DOI: 10.1109/acssc.2013.6810313. 5, 15, 17

[86] S. Vu, C. T. Gao, W. P. Deshmukh, and L. Yingshu, Distributed energy-efficient scheduling approach for k-coverage in wireless sensor networks, *IEEE Military Communications Conference*, pp. 1–7, October 2006. DOI: 10.1109/milcom.2006.302146. 5, 6, 39

[87] J. A. Deri and J. Moura, Graph sampling: Estimation of degree distributions, *IEEE International Conference on Acoustics, Speech and Signal Processing (ICASSP)*, pp. 6501–6505, May 2013. DOI: 10.1109/icassp.2013.6638918. 6, 7

[88] D. Rojković, T. Crnić, and I. Cavrak, Agent-based topology control for wireless sensor network applications, *Proc. of the 35th International Convention MIPRO*, pp. 277–282, May 2012. 5, 6

[89] O. Serrat, *Social Network Analysis*, Singapore, Springer Singapore, pp. 39–43, 2017. DOI: 10.1007/978-981-10-0983-9_9. 5

[90] D. W. Hearn and J. Vijay, Efficient algorithms for the (weighted) minimum circle problem, *Operations Research*, 30, pp. 777–795, July 1982. DOI: 10.1287/opre.30.4.777. 6, 39

[91] Z. Yu, J. Teng, X. Li, and D. Xuan, On wireless network coverage in bounded areas, *INFOCOM, Proceedings IEEE*, pp. 1195–1203, July 2013. DOI: 10.1109/infcom.2013.6566911. 6, 39

[92] Z. Sun, P. Wang, M. Vuran, M. Al-Rodhaan, A. Al-Dhelaan, and I. Akyildiz, Bordersense: Border patrol through advanced wireless sensor networks, *Ad Hoc Networks*, 9, pp. 468–477, May 2011. DOI: 10.1016/j.adhoc.2010.09.008. 6

[93] N. M. M. de Abreu, Old and new results on algebraic connectivity of graphs, *Linear Algebra and its Applications*, 423, pp. 53–73, May 2007. DOI: 10.1016/j.laa.2006.08.017. 9, 17

[94] S. Lu and Z. Wang, Accelerated algorithms for eigen-value decomposition with application to spectral clustering, *49th Asilomar Conference on Signals, Systems and Computers*, pp. 355–359, November 2015. DOI: 10.1109/acssc.2015.7421146. 10, 17, 56

[95] P. D. Lorenzo and S. Barbarossa, Distributed estimation and control of algebraic connectivity over random graphs, *IEEE Transactions on Signal Processing*, 62(21), pp. 5615–5628, November 2014. DOI: 10.1109/tsp.2014.2355778. 10, 17

[96] A. Nedic, A. Ozdaglar, and P. A. Parrilo, Constrained consensus and optimization in multi-agent networks, *IEEE Transactions on Automatic Control*, 55(4), pp. 922–938, April 2010. DOI: 10.1109/tac.2010.2041686. 11

[97] D. P. Spanos and R. M. Murray, Distributed sensor fusion using dynamic consensus, *IFAC World Congress*, pp. 1–6, July 2005. 11

[98] D. Scherber and H. Papadopoulos, Locally constructed algorithms for distributed computations in ad-hoc networks, *3rd International Symposium on Information Processing in Sensor Networks*, pp. 11–19, June 2004. DOI: 10.1145/984622.984625. 11

[99] L. Schenato and G. Gamba, A distributed consensus protocol for clock synchronization in wireless sensor network, *46th IEEE Conference on Decision and Control*, pp. 2289–2294, December 2007. DOI: 10.1109/cdc.2007.4434671. 11

[100] S. Kar, J. M. F. Moura, and K. Ramanan, Distributed parameter estimation in sensor networks: Nonlinear observation models and imperfect communication, *IEEE Transactions on Information Theory*, 58(6), pp. 3575–3605, June 2012. DOI: 10.1109/tit.2012.2191450. 11

[101] I. D. Schizas, G. B. Giannakis, S. I. Roumeliotis, and A. Ribeiro, Consensus in ad hoc wsns with noisy links—Part II: Distributed estimation and smoothing of random signals, *IEEE Transactions on Signal Processing*, 56(4), pp. 1650–1666, April 2008. DOI: 10.1109/TSP.2007.908943.

[102] V. Saligrama, M. Alanyali, and O. Savas, Distributed detection in sensor networks with packet losses and finite capacity links, *IEEE Transactions on Signal Processing*, 54(11), pp. 4118–4132, November 2006. DOI: 10.1109/tsp.2006.880227. 11

[103] M. Bawa, H. Garcia-Molina, A. Gionis, and R. Motwani, Estimating aggregates on a peer-to-peer network, Stanford Info Lab, Technical Report 2003–24, April 2003. `http://ilpubs.stanford.edu:8090/586/` 11, 17, 18

[104] S. Sundaram and C. N. Hadjicostis, Distributed function calculation via linear iterative strategies in the presence of malicious agents, *IEEE Transactions on Automatic Control*, 56(7), pp. 1495–1508, July 2011. DOI: 10.1109/tac.2010.2088690. 11

[105] A. Jadbabaie, J. Lin, and A. Morse, Coordination of groups of mobile autonomous agents using nearest neighbor rules, *IEEE Transactions on Automatic Control*, 48, pp. 988–1001, June 2003. DOI: 10.1109/cdc.2002.1184304. 11

[106] X. Li, C. Tepedelenlioglu, and H. Şenol, Channel estimation for residual self-interference in full-duplex amplify-and-forward two-way relays, *IEEE Transactions on Wireless Communications*, 16(8), pp. 4970–4983, August 2017. DOI: 10.1109/twc.2017.2704123. 12

[107] J. Cortes, Distributed algorithms for reaching consensus on general functions, *Automatica*, 44(3), pp. 401–420, 2008. DOI: 10.1016/j.automatica.2007.07.022. 14, 15

[108] D. Bauso, L. Giarre, and R. Pesenti, Nonlinear protocols for optimal distributed consensus in networks of dynamic agents, *System and Control Letters*, 55(11), pp. 918–928, June 2006. DOI: 10.1016/j.sysconle.2006.06.005. 15

[109] S. Zhang, S. C. Liew, and P. P. Lam, Hot topic: Physical-layer network coding, *Proc. of the 12th Annual International Conference on Mobile Computing and Networking*, pp. 358–365, September 2006. DOI: 10.1145/1161089.1161129. 16

[110] J. Sykora, Hierarchical network transfer function and doubly-greedy half-duplex stage scheduling for WPNC networks, *IEEE Communications Letters*, 19(6), pp. 1029–1032, June 2015. DOI: 10.1109/lcomm.2015.2417874. 16

[111] A. Ganesh, A. Kermarrec, E. Merrer, and L. Massoulie, Peer counting and sampling in overlay networks based on random walks, *Distributed Computing*, 20, pp. 267–278, January 2007. DOI: 10.1007/s00446-007-0027-z. 16, 17, 18

[112] J. Sykora, Distributed consensus estimator of hierarchical network transfer function in WPNC networks, *COST IRACON*, pp. 1–4, October 2016. 16, 17

[113] X. Zhang, The Laplacian eigenvalues of graphs: A survey, *arXiv:1111.2897v1 [math.CO]*, pp. 1–35, November 2011. 17

[114] P. Yang, F. R. A., G. G. J., K. Jynch, S. Srinivasa, and R. Sukthankar, Decentralized estimation and control of graph connectivity for mobile sensor networks, *Automatica*, 46(2), pp. 390–396, February 2010. DOI: 10.1016/j.automatica.2009.11.012. 17

[115] W. Y. Chen, Y. Song, H. Bai, C. J. Lin, and E. Y. Chang, Parallel spectral clustering in distributed systems, *IEEE Transactions on Pattern Analysis and Machine Intelligence*, 33(3), pp. 568–586, March 2011. DOI: 10.1109/tpami.2010.88. 17

[116] M. Talistu, T. S. Moh, and M. Moh, Gossip-based spectral clustering of distributed data streams, *International Conference on High Performance Computing Simulation (HPCS)*, pp. 325–333, July 2015. DOI: 10.1109/hpcsim.2015.7237058. 17

[117] R. Zeng and C. Tepedelenlioglu, Underlay cognitive multiuser diversity with random number of secondary users, *IEEE Transactions on Wireless Communications*, 13(10), pp. 5571–5581, October 2014. DOI: 10.1109/twc.2014.2340868. 17

[118] R. Zeng and C. Tepedelenlioglu, Fundamental BER performance trade-off in cooperative cognitive radio systems with random number of secondary users, *50th Asilomar Conference on Signals, Systems and Computers*, pp. 889–893, November 2016. DOI: 10.1109/acssc.2016.7869177.

[119] R. Zeng and C. Tepedelenlioglu, Fundamental performance trade-offs in cooperative cognitive radio systems, *IEEE Transactions on Cognitive Communications and Networking*, 3(2), pp. 169–179, June 2017. DOI: 10.1109/tccn.2017.2701814. 17

[120] B. Ribeiro and D. Towsley, Estimating and sampling graphs with multidimensional random walks, *Proc. of the 10th Annual Conference on Internet Measurement*, pp. 390–403, November 2010. DOI: 10.1145/1879141.1879192. 18

[121] C. Gkantsidis, M. Mihail, and A. Saberi, Random walks in peer-to-peer networks: Algorithms and evaluation, *Performance Evaluation*, 63, pp. 241–263, March 2006. DOI: 10.1016/j.peva.2005.01.002. 18

[122] R. Ali, S. Lor, and M. Rio, Two algorithms for network size estimation for master/slave ad hoc networks, *arXiv:0908*, 56, October 2009. DOI: 10.1109/ants.2009.5409896.

[123] D. Kostoulas, D. Psaltoulis, I. Gupta, K. Birman, and A. Demers, Active and passive techniques for group size estimation in large-scale and dynamic distributed systems, *Journal of Systems and Software*, 80, pp. 1639–1658, October 2007. DOI: 10.1016/j.jss.2007.01.014.

[124] K. Horowitz and D. Malkhi, Estimating network size from local information, *Information Processing Letters*, 88, pp. 237–243, December 2003. DOI: 10.1016/j.ipl.2003.08.011.

[125] S. Peng, S. Li, X. Liao, Y. Peng, and N. Xiao, Estimation of a population size in large-scale wireless sensor networks, *Journal of Computer Science and Technology*, 24, pp. 987–997, September 2009. DOI: 10.1007/s11390-009-9273-9. 18

[126] D. Saha and P. S. Das, Nabanita, *A Digital-Geometric Approach for Computing Area Coverage in Wireless Sensor Networks*, Lecture Notes in Computer Science, book series, Springer International Publishing Switzerland, 2014. DOI: 10.1007/978-3-319-04483-5_15. 18, 19

[127] E. Le Merrer, A.-M. Kermarrec, and L. Massoulie, Peer to peer size estimation in large and dynamic networks: A comparative study, *15th IEEE International Symposium on High Performance Distributed Computing*, pp. 7–17, 2006. DOI: 10.1109/hpdc.2006.1652131. 18

[128] D. Kostoulas, D. Psaltoulis, I. Gupta, K. Birman, and K. Demers, Decentralized schemes for size estimation in large and dynamic groups, *4th IEEE International Symposium on Network Computing and Applications*, pp. 41–48, July 2005. DOI: 10.1109/nca.2005.15. 18

[129] G. S. Manku, M. Bawa, and P. Raghavan, Symphony: Distributed hashing in a small world, *USITS'03 Proceedings of the 4th Conference on USENIX Symposium on Internet Technologies and Systems*, 2003. 18

[130] D. Varagnolo, G. Pillonetto, and L. Schenato, Distributed cardinality estimation in anonymous networks, *IEEE Transaction on Automatic Control*, 59(3), pp. 645–659, March 2014. DOI: 10.1109/tac.2013.2287113. 18

[131] D. Varagnolo, G. Pillonetto, and L. Schenato, Distributed statistical estimation of number of nodes in networks, *49th IEEE Conference on Decision and Control*, pp. 1498–1503, December 2010. DOI: 10.1109/cdc.2010.5717355. 18

[132] S. Kamath, D. Manjunath, and R. Mazumdar, On distributed function computation in structure-free random wireless networks, *IEEE Transactions on Information Theory*, 60(1), pp. 432–442, January 2014. DOI: 10.1109/tit.2013.2293214.

[133] H. Terelius, D. Varagnolo, and K. H. Johansson, Distributed size estimation of dynamic anonymous networks, *51st IEEE Conference on Decision and Control*, pp. 5221–5227, December 2012. DOI: 10.1109/cdc.2012.6425912. 18

[134] I. Shames, T. Charalambous, C. N. Hadjicostis, and M. Johansson, Distributed network size estimation and average degree estimation and control in networks isomorphic to directed graphs, *50th Annual Allerton Conference*, pp. 1885–1892, October 2012. DOI: 10.1109/allerton.2012.6483452. 18

[135] M. Jelasity and A. Montresor, Epidemic-style proactive aggregation in large overlay networks, *Proc. of the 24th International Conference on Distributed Computing Systems*, pp. 102–109, 2004. DOI: 10.1109/icdcs.2004.1281573. 18

[136] S. Zhang, *Consensus Algorithms and Distributed Structure Estimation in Wireless Sensor Networks*, Ph.D. Dissertation, Arizona State University, May 2017. 18, 39, 41

[137] S. Kundu and N. Das, In-network area estimation and localization in wireless sensor networks, *The 7th IEEE International Workshop on Heterogeneous, Multi-hop, Wireless and Mobile Networks*, pp. 431–435, 2012. DOI: 10.1109/glocomw.2012.6477611. 18, 19

[138] S. Li, H. Fan, and Y. Wang, Finding the smallest ellipse containing a point set based on genetic algorithms, *IEEE International Symposium on Knowledge Acquisition and Modeling Workshop*, pp. 693–696, December 2008. DOI: 10.1109/kamw.2008.4810584. 18, 19

[139] B. Greenstein, E. Kohler, D. Culler, and D. Estrin, Distributed techniques for area computation in sensor networks, *29th Annual IEEE International Conference on Local Computer Networks*, pp. 1–9, November 2004. DOI: 10.1109/lcn.2004.45. 18

[140] S. Boyd and L. Vandenberghe, *Convex Optimization*, Cambridge University Press, 2004. DOI: 10.1017/cbo9780511804441. 18, 19, 41

[141] S. Zhang, C. Tepedelenlioglu, M. Banavar, and A. Spanias, Distributed node counting in wireless sensor networks in the presence of communication noise, *IEEE Sensors Journal*, 17, pp. 1175–1186, February 2017. DOI: 10.1109/jsen.2016.2640943. 21, 23, 25

[142] S. Zhang, C. Tepedelenlioglu, M. Banavar, and A. Spanias, Distributed node counting in wireless sensor networks, *49th Asilomar Conference on Signals Systems and Computers*, November 2015. DOI: 10.1109/acssc.2015.7421147. 25

[143] S. Zhang, C. Tepedelenlioglu, J. Lee, H. Braun, and A. Spanias, Cramer-Rao bounds for distributed system size estimation using consensus algorithms, *Sensor Signal Processing for Defence*, pp. 1–5, Edinburgh, September 2016. DOI: 10.1109/sspd.2016.7590591. 21

[144] C. Bettstetter, On the minimum node degree and connectivity of a wireless multihop network, *Proc. of the 3rd ACM International Symposium on Mobile ad hoc Networking and Computing*, pp. 80–91, 2002. DOI: 10.1145/513800.513811. 29

[145] C. Bettstetter, J. Klinglmayr, and S. Lettner, On the degree distribution of k-connected random networks, *Proc. IEEE International Conference on Communications (ICC)*, pp. 1–6, May 2010. DOI: 10.1109/icc.2010.5502272. 29

[146] M. Newman, The structure and function of complex networks, *SIAM Review*, 45(2), pp. 167–256, January 2003. DOI: 10.1137/s003614450342480. 29, 32

[147] S. Zhang, J. Lee, C. Tepedelenlioglu, and A. Spanias, Distributed estimation of the degree distribution in wireless sensor networks, *IEEE Global Communications Conference (GLOBECOM)*, pp. 1–6, December 2016. DOI: 10.1109/glocom.2016.7841740. 29, 34

[148] S. Zhang, C. Tepedelenlioglu, and A. Spanias, Distributed center and coverage region estimation in wireless sensor networks using diffusion adaptation, *Asilomar Conference on Signals, Systems and Computers*, 2017 (accepted). 39

[149] S. Zhang, C. Tepedelenlioglu, and A. Spanias, Distributed network center and size estimation, *IEEE Sensors Journal*, 2017 (submitted). 39

[150] G. Montavon, G. B. Orr, and K. R. Muller, *Neural Networks: Tricks of the Trade*, Lecture Notes in Computer Science, book series, Springer Berlin Heidelberg, 2012. DOI: 10.1007/978-3-642-35289-8. 41

[151] J. Chen and A. H. Sayed, Diffusion adaptation strategies for distributed optimization and learning over networks, *IEEE Transactions on Signal Processing*, 60, pp. 4289–4305, August 2012. DOI: 10.1109/tsp.2012.2198470. 42, 43, 45

[152] A. Nemirovski, A. Juditsky, G. Lan, and A. Shapiro, Robust stochastic approximation approach to stochastic programming, *SIAM Journal of Optimization*, 19(4), pp. 1574–1609, January 2009. DOI: 10.1137/070704277. 45

[153] M. Xiang, L. Sun, and L. Li, Survey on the connectivity and coverage in wireless sensor networks, *7th International Conference on Wireless Communications, Networking and Mobile Computing*, pp. 1–4, September 2011. DOI: 10.1109/wicom.2011.6040351. 56

[154] G. Zhi-yan and W. Jian-zhen, Research on coverage and connectivity for heterogeneous wireless sensor network, *7th International Conference on Computer Science Education (ICCSE)*, pp. 1239–1242, July 2012. DOI: 10.1109/iccse.2012.6295289.

[155] W. Qihua, G. Ge, C. Lijie, and X. Xufeng, Voronoi coverage algorithm based on connectivity for wireless sensor networks, *34th Chinese Control Conference (CCC)*, pp. 7833–7837, July 2015. DOI: 10.1109/chicc.2015.7260884. 56

[156] Y. Zhou, A. Ortega, D. Wang, and S. Lee, Node clustering for data collection in wireless sensor networks using graph-transform and compressive sampling, *12th International Conference on Signal Processing (ICSP)*, pp. 2251–2256, October 2014. DOI: 10.1109/icosp.2014.7015395. 56

[157] R. Bhatt and R. Datta, Utilizing graph sampling and connected dominating set for backbone construction in wireless multimedia sensor networks, *20th National Conference on Communications (NCC)*, pp. 1–6, February 2014. DOI: 10.1109/ncc.2014.6811362.

[158] H. Zheng, F. Yang, X. Tian, X. Gan, X. Wang, and S. Xiao, Data gathering with compressive sensing in wireless sensor networks: A random walk based approach, *IEEE Transactions on Parallel and Distributed Systems*, 26(1), pp. 35–44, January 2015. DOI: 10.1109/tpds.2014.2308212. 56

[159] J. Zhou, L. Chen, C. L. P. Chen, Y. Wang, and H. X. Li, Uncertain data clustering in distributed peer-to-peer networks, *IEEE Transactions on Neural Networks and Learning Systems*, PP(99), pp. 1–15, 2017. DOI: 10.1109/tnnls.2017.2677093. 56

[160] J. Qin, W. Fu, H. Gao, and W. X. Zheng, Distributed k-means algorithm and fuzzy c-means algorithm for sensor networks based on multiagent consensus theory, *IEEE Transactions on Cybernetics*, 47(3), pp. 772–783, March 2017. DOI: 10.1109/tcyb.2016.2526683. 56

[161] M. A. Alsheikh, S. Lin, D. Niyato, and H. P. Tan, Machine learning in wireless sensor networks: Algorithms, strategies, and applications, *IEEE Communications Surveys Tutorials*, 16(4), pp. 1996–2018, 2014. DOI: 10.1109/comst.2014.2320099. 56

[162] T. Shafaat, A. Ghodsi, and S. Haridi, *A Practical Approach to Network Size Estimation for Structured Overlays*, Lecture Notes in Computer Science, book series (LNCS, volume 5343), 2008. DOI: 10.1007/978-3-540-92157-8_7.

[163] A. Forster and A. Murphy, *Machine Learning Across the WSN Layers* InTech, 2011. DOI: 10.5772/10516.

[164] Y. Zhang, N. Meratnia, and P. Havinga, Outlier detection techniques for wireless sensor networks: A survey, *IEEE Communications Surveys Tutorials*, 12(2), pp. 159–170, 2010. DOI: 10.1109/surv.2010.021510.00088.

[165] A. Wisler, V. Berisha, A. Spanias, and A. Hero, Direct estimation of density functionals using a polynomial basis, *IEEE Transactions on Signal Processing*, 66(3), pp. 558–572, 2017. DOI: 10.1109/TSP.2017.2775587.

[166] V. Berisha, A. Wisler, A. O. Hero, and A. Spanias, Empirically estimable classification bounds based on a nonparametric divergence measure, *IEEE Transactions on Signal Processing*, 64(3), pp. 580–591, February 2016. DOI: 10.1109/tsp.2015.2477805.

[167] A. Forster and A. L. Murphy, Clique: Role-free clustering with q-learning for wireless sensor networks, *29th IEEE International Conference on Distributed Computing Systems*, pp. 441–449, June 2009. DOI: 10.1109/icdcs.2009.43.

[168] M. Mihaylov, K. Tuyls, and A. Nowé, *Decentralized Learning in Wireless Sensor Networks*, pp. 60–73, Berlin, Heidelberg, Springer Berlin Heidelberg, 2010. DOI: 10.1007/978-3-642-11814-2_4. 56

Authors' Biographies

SAI ZHANG

Sai Zhang received a B.S. degree in electrical and information engineering from Huazhong University of Science and Technology, Wuhan, China, in 2012 and an M.S. degree in electrical engineering from Arizona State University, Tempe, AZ, in 2014. From 2014 to 2017 he was a research assistant at Arizona State University, where he completed his Ph.D. degree in electrical engineering. His research interests include distributed computation in wireless sensor networks, performance analysis of distributed consensus algorithms, and wireless communications.

CIHAN TEPEDELENLIOGLU

Cihan Tepedelenlioglu was born in Ankara, Turkey, in 1973. He received his B.S. degree with highest honors from Florida Institute of Technology in 1995, and his M.S. degree from the University of Virginia in 1998, both in Electrical Engineering. From January 1999 to May 2001 he was a research assistant at the University of Minnesota, where he completed his Ph.D. degree in Electrical and Computer Engineering. He is currently an Associate Professor of Electrical Engineering at Arizona State University. He was awarded the NSF (early) Career grant in 2001 and has served as an Associate Editor for several IEEE Transactions including *IEEE Transactions on Communications*, *IEEE Signal Processing Letters*, and *IEEE Transactions on Vehicular Technology*.

His research interests include statistical signal processing, system identification, wireless communications, estimation and equalization algorithms for wireless systems, multiantenna communications, OFDM, ultra-wideband systems, distributed detection and estimation, and data mining for photovoltaic systems.

ANDREAS SPANIAS

Andreas Spanias is Professor in the School of Electrical, Computer, and Energy Engineering at Arizona State University. He is also the founder and director of the SenSIP Industry Consortium. His research interests are in the areas of adaptive signal processing, speech processing, and audio sensing. He and his student team developed the computer simulation software Java-DSP (J-DSP - ISBN 0-9724984-0-0). He is author of two textbooks: *Audio Processing and Coding* by Wiley and *DSP: An Interactive Approach*. He served as Associate Editor of the IEEE Transactions on Signal Processing and as General Co-Chair of IEEE ICASSP-99. He also served as the IEEE Signal Processing Vice President for Conferences. Andreas Spanias is co-recipient of the 2002 IEEE Donald G. Fink paper prize award and was elected Fellow of the IEEE in 2003. He served as Distinguished Lecturer for the IEEE Signal Processing Society in 2004.

MAHESH BANAVAR

Mahesh Banavar received a B.E. degree in telecommunications engineering from Visvesvaraya Technological University, Karnataka, India, in 2005 and M.S. and Ph.D. degrees, both in electrical engineering, from Arizona State University, Tempe, in 2007 and 2010, respectively. He is currently an Assistant Professor in the Department of Electrical and Computer Engineering at Clarkson University, Potsdam, NY. His interests include node localization, detection and estimation algorithms, and performance analysis of distributed sensor algorithms for wireless sensor networks. Dr. Banavar is a recipient of the Teaching Excellence Award from the Graduate and Professional Student Association at Arizona State University and the Outstanding Teaching Award from the Eta Kappa Nu chapter at Clarkson University. He is also a member of MENSA and the Eta Kappa Nu honor society.